Studies in Systems, Decision and Control

Volume 638

The series "Studies in Systems, Decision and Control" (SSDC) covers both new developments and advances, as well as the state of the art, in the various areas of broadly perceived systems, decision making and control–quickly, up to date and with a high quality. The intent is to cover the theory, applications, and perspectives on the state of the art and future developments relevant to systems, decision making, control, complex processes and related areas, as embedded in the fields of engineering, computer science, physics, economics, social and life sciences, as well as the paradigms and methodologies behind them. The series contains monographs, textbooks, lecture notes and edited volumes in systems, decision making and control spanning the areas of Cyber-Physical Systems, Autonomous Systems, Sensor Networks, Control Systems, Energy Systems, Automotive Systems, Biological Systems, Vehicular Networking and Connected Vehicles, Aerospace Systems, Automation, Manufacturing, Smart Grids, Nonlinear Systems, Power Systems, Robotics, Social Systems, Economic Systems and other. Of particular value to both the contributors and the readership are the short publication timeframe and the world-wide distribution and exposure which enable both a wide and rapid dissemination of research output.

Indexed by SCOPUS, DBLP, WTI Frankfurt eG, zbMATH, SCImago.

All books published in the series are submitted for consideration in Web of Science.

Rimantas Pupeikis · Kazys Kazlauskas

Advanced Digital Signal Processing Methods for Filtering, Identification, and Nonlinear Systems Control

Springer

Rimantas Pupeikis
Faculty of Mathematics and Informatics
Vilnius University
Vilnius, Lithuania

Kazys Kazlauskas
Faculty of Mathematics and Informatics
Vilnius University
Vilnius, Lithuania

ISSN 2198-4182 ISSN 2198-4190 (electronic)
Studies in Systems, Decision and Control
ISBN 978-3-032-06739-5 ISBN 978-3-032-06740-1 (eBook)
https://doi.org/10.1007/978-3-032-06740-1

This Springer imprint is published by the registered company Springer Nature Switzerland AG
The registered company address is: Gewerbestrasse 11, 6330 Cham, Switzerland

Preface

The book, named *Advanced Methods for Digital Systems and Signal Processing*, published by Shaker, contains models and approaches of analysis and calculation of their main characteristics, as well as methods of filtering and parametrical identification of systems (filters). The content of the previous book has been complemented by this one, where methods of analysis, filtering, and identification of adaptive nonlinear systems proposed for minimum variance (MV) servo control are presented. The set of monographs' started in the book (Kazlauskas and Pupeikis 2016), is updated by this one and is to be very likely continued. A great stimulus of the exploration was got in the succesful start of the New Visby project Ref. No. 2473/2002 (381/T81) of the Royal Swedish Academy of Sciences and the Swedish Institute (chief of the project Prof. D. Navakauskas) during the long-time visit at the Division of Automatic Control of Linköping University (head of the control group Prof. L. Ljung). Nowadays, it took the acceleration at the Institute of Data Science and Digital Technologies of the Faculty of Mathematics and Informatics of Vilnius University, where we continued our scientific activity.

The objective of this monograph is original methods and algorithms for adaptive filtering and parametrical identification of Volterra, polynomial autoregressive (AR), bilinear, and, partly, switched systems. The MV control approach has been developed using well-known minimum variance or generalized minimum variance (MV/GMV) self-tuning controllers (STC) for the block-oriented systems having piecewise-linear (PWL) nonlinearities. Controllers are based on the numeric minimization of the cost function, i.e., the variance of the error between a set-point and the controlled output k-time steps in the future signals. The desired controller is the one that minimizes this variance, hence the name MV control.

Like the preceding creations, this monograph is a result of approximately 30-year experience of the authors in the research on adaptive filters and systems as well on their identification-based optimization for adaptive control. The roots of it started at the former Laboratory of Adaptive systems at Institute of Physical and Technical Problems of Energetics in Kaunas where we have been taken usefull skills. The experience in scientific research we have been gather in the Group of Control of Technological Processes at Process Recognition Department of the Mathematics and Informatics Institute in Vilnius. Teaching of informatics at the Vilnius Educational Science University in the Department of Informatics, and the digital signal processing (DSP) at the Department of Electronics systems of Electronics Faculty of Vilnius Gediminas Technical (Vilnius Tech) University stimulated fast uprise of our experience. Collaborative investigations skills have been sharpened during the development and execution of the Deutsche Forshungsgemeinshaft project "Echtzeit-Optimierung der Grössen Systemen" (chief of the project Prof. P. J. Huber), as well as abovementioned New Visby project at the Division of Automatic Control of Linköping University. Collaborative investigations rendered active scientific results that were represented by one of authors at DFG workshops in Hof (1996) and Bad Fredeburg (1997), and at the seminar of Automatic Control Division of Linchöping University (2002), as well as at the European Conference on Circuit Theory and Design in Cracow (2003). Most of further results were published in local journals included in Science Citation Index Expanded (SciSearch, available at http://scientic.thomson.com), such as "Informatica" (the journal of Mathematics and Informatics Institute) and "Information Technology and Control" (the journal of Kaunas University of Technology), as well as in the worldwide international journals: "Robust and Nonlinear Control"(2014), and "Adaptive Control and Signal Processing" (2020, 2021).

The monograph is aimed at three major groups of readers: senior undergraduate students, graduate students, and scientific research workers in electrical engineering, computer engineering, computer science, and digital control. The monograph could be suitable for researchers as the background reading and references, as well.

The monograph consists of nine chapters.

The text begins with an introduction to the monograph, where the reasons of its publishing are shown and explained briefly. Here the basic approaches to the problems, to be analyzed in the book, are outlined and expressed in short. Many but, regrettably, limited references to the well-known and recently published works are given.

The well-known least mean square (LMS) algorithm is explored theoretically in Chap. 2. It is speeded up without changing its adaptive characteristics. A high speed is guaranteed by increasing the parallelism in the LMS adaptive algorithm. Iterative procedures are given for efficient computation of the lower triangular inverse matrix and the input signal covariance matrix, as well.

In Chap. 3, the block algorithm is proposed for optimal estimation of parameters of the linear periodically time-variant (LPTV) system, using the samples of input-output signals, acting in a noisy frame. The simulation results are presented that corroborate the efficiency of the approach.

Chapter 4 represents an iterative autoregressive system parameter estimation algorithm in the presence of additive white noise. To eliminate the parameter estimation bias, the original bias correction approach is used. Simulation results, based on the Monte Carlo experiments, are presented, too.

Chapter 5 shows that nonlinear Volterra, polynomial AR and bilinear systems have the same layered implementation routine. Using the layered structure, the order of nonlinearity can be increased/decreased by adding/deleting more layers to/from the structure. In addition, the modular layered structures admit the very large-scale integration (VLSI) implementation of the polynomial nonlinear filters.

Chapter 6 introduces potential linear time-invariant (LTI) system models in the state-space and in the time-domain. It is shown here that LTI systems in both spaces can be described by the well-known autoregressive moving average with exogenous input (ARMAX) models. Methods and algorithms for parametrical identification of LTI systems having piecewise-linear (PWL) nonlinearities are described here, as well. The system itself consists of two blocks: a linear dynamic block and a nonlinear static one. It is known that the PWL nonlinearity can be modeled as the regression function in "pieces" (see https://www.online/stat.psu.edu/stat501/lesson/8/). On the other hand, such a nonlinearity belongs to the class of switched PWL nonlinear functions. There exists opinion that LTI systems having the PWL nonlinearities, as follows from the cited literature, can be controlled efficiently, using the MV/GMV control law. Flowcharts are realized, by means of ordinary self-tuning MV controllers, based on the optimization of the applicable quadratic MV criteria. The servo control simulation results in a noisy frame are shown and discussed here, too.

Chapter 7 analyzes a general approach, used for a parametric identification of block-oriented Wiener systems, as well as for the Hammerstein ones with hard PWL nonlinearities. To separate exactly different segments of the PWL nonlinearity it is used the data reordering in an ascending order of their sampling values. Before the parametric identification of the nonlinear system the threshold values were determined. An original technique is proposed here for the estimation of parameters of linear and nonlinear parts of a system, using reordered data samples. The simulation results for open-loop linear time-variant (LTV) systems with invertible and non-invertible PWL nonlinearities are given.

Chapter 8 considers the parametrical identification of LTV systems, followed or preceded by the deadzone. The same approach, based on the reodering of observations in an ascending order of their sample values, is used. Numerical Monte Carlo simulation results are obtained by parametrical identification of Wiener and Hammerstein systems.

Chapter 9 has set a great deal on adaptive control of Wiener systems, containing switched PWL nonlinearities and operating in a noisy environment. The information, needed for this type of servo control, is obtained during the learning (*training*) phase based on the reordered data, which can be used for invertible, as well as for noninvertible PWL nonlinearities. The simulation results of control of Wiener systems are given, using input-noisy output observations, stored in advance.

Appendix represents the estimates consistency results shortly and multifold well-known examples of the use of DSP methods for modelling, filtering, identification, and control.

Chapters 2–5 are written by K. Kazlauskas while Chaps. 6–9 are written by R. Pupeikis. The introduction is written jointly.

The authors thank the Institute of Data Science and Digital Technologies of Vilnius University as well as the Faculty of Mathematics and Informatics of Vilnius University for a long-term financial support of their scientific researches.

Vilnius, Lithuania

Rimantas Pupeikis
Kazys Kazlauskas

Competing Interests The authors have no competing interests to declare that are relevant to the content of this manuscript.

Contents

Acronyms

AR	Autoregressive
ARMA	Autoregressive moving average
ARMAX	Autoregressive moving average with exogenous input
ARX	Autoregressive with exogenous input
DCT	Discrete cosine transformation
DSP	Digital signal processing
FIR	Finite impulse response
FPGA	Field programmable gate array
GMV	Generalized minimum variance
GMVC	Generalized minimum variance control
HO(Y-W)	High-order Yule-Walker
ILSD	Improved least squares with direct implementation structure
ILSPN	Improved least squares with no prefiltering
LMS	Least mean square
LP	Linear prediction
LPTV	Linear periodically time-variant
LS	Least squares
LTI	Linear time-invariant
LTV	Linear time-variant
MA	Moving-average
MFR	Magnitude-frequency response
MV	Minimum variance
MV/GMV	Minimum variance or generalized MV
MVC	Minimum variance control
PAR	Polynomial AR
PLR	Pseudo-linear regression
PTV	Periodically time-variant
PWA	Piecewise-affine
PWL	Piecewise-linear
RGLS	Recursive generalized least squares
RLS	Recursive least squares

RML	Recursive maximum likelihood
SISO	Single-input single-output
SNR	Signal-to-noise ratio
ST	Self-tuning
STC	Self-tuning controller
std	Standard deviation
VLSI	Very large scale integration
Y-W	Yule-Walker
$\mathcal{E}$	Mathematical expectation

Chapter 1
Introduction

A number of researches, devoted to the computer based analysis of nonlinear effects in system identification and adaptive control of such systems, are continuously increasing over a time. Some problems were reviewed also in our previous monographies: Kazlauskas and Pupeikis 2016; [84, 86]. The investigations were continued. Some actual problems have been explored. We begin here the monograph with the examination of the adaptive filters (Chap. 2), then, continuing about linear models, including the estimation of parameters of LPTV systems and AR processes (Chaps. 3 and 4), and examining the layered polynomial filter structures, used for nonlinear filters (Chap. 5). The next part of the book deals with the exploration of potentiality for the possibility of adaptive control of linear ARMAX systems with PWL nonlinearities (Chap. 6). The original approach, based on the linear block convolution, and used for reordering of the stored data for parametrical identification of nonlinear Hammerstein and Wiener systems having three or more segment PWL nonlinearities (Chaps. 7 and 8) is explored. It gives us a possibility to identify block-oriented systems with the invertible PWL nonlinearity as well as with the noninvertible one parametrically. It allows us for the noninvertible nonlinearity to build in the training phase the one-shot MV controller tuning, i.e. parametrical identification and controller parameter calculation routines are connected jointly only during the training time. Chapter 9 includes the extended computer-based experimental simulation and examination of MV/GMV control in a noisy frame.

One of the main problems is realization of adaptive algorithms as well as filters at high sampling rates. Adaptive algorithms (filters), based on the transversal filters, have an inherent sampling rate limitation for a given speed of hardware due to the feedback of the residual error in the adaptation of individual stages [66].

In Chap. 2, an approach is proposed for speeding up the popular LMS algorithm without changing any of its adaptive characteristics. The time-domain LMS adaptive filter has been used in a variety of applications such as system modelling, noise cancelling, channel equalization, adaptive antenna, and adaptive array processing [50, 118, 119]. In the usual implementation, the outputs of a tapped delay line

R. Pupeikis and K. Kazlauskas, *Advanced Digital Signal Processing Methods for Filtering, Identification, and Nonlinear Systems Control*, Studies in Systems, Decision and Control 638, https://doi.org/10.1007/978-3-032-06740-1_1

are weighted and summed to produce the filter output. The weights of this transversal filter which determines its impulse response, are updated iteratively, basing on the difference between the filter output and a desired input, so as to minimize the mean-square error of the difference. The frequency domain implementation of an adaptive filter, requiring less computation than its time domain counterpart, has been proposed by [26]. In some applications, a good performance can still be obtained by frequency domain techniques based on circular convolution, e.g., in the detection of narrow-band signals in broad-band noise [33]. Due to the availability of high speed parallel processors and fast Fourier transform algorithms for convolution, block digital filtering has got great attention [69, 70, 110]. It is shown that this drawback can be overcome, and that the availability of block processing techniques allows the derivation of computationally efficient parallel algorithms that behave exactly just like the same way as their scalar version. These algorithms are just a rearrangement of the initial LMS equations. We propose a realization of the conventional LMS adaptive filter in the time domain which does converge to the optimum transversal filter solution. The proposed method is intended to be a direct replacement of the LMS adaptive filter, but requires less computation than the time domain realization. The proposed algorithm converges at the same rate as the time domain LMS filter. Convergence at this rate is not always advantageous, e.g., when the correlation matrix of the signals at the filter taps has highly disparate eigenvalues. Nevertheless, the LMS adaptive filter has been widely used and is well understood, so that a fast implementation should be of interest.

In Chap. 3 the estimation of the parameters of LPTV systems are presented. It is known that the problem of estimating parameters by noisy observed data has a long history in engineering and experimental science. These problems have been analysed in a large number of works in many different fields of system modelling and identification [31, 53, 54, 63, 64]. Among many models, the ARMA model is one of the most effective and practical ones. When the observations and the unknown parameters are related by a linear model, and a stochastic setting is assumed, then the application of the maximum likelihood principle [7] leads to the well-known LS parameter estimate. The main reason for the popularity of ARMA modelling is that it is possible to obtain good estimates of the unknown parameters by solving a simultaneous set of linear equations. Many well-known techniques, developed for statistical time series analysis, have been used successfully in the deterministic case [55]. An LPTV discrete-time system with period L is a system for which a shift in the input sequence by L samples results in a shift of L samples in the output sequence. These systems are a generalization of LTI systems and are frequently used in control systems, signal processing, and digital communications [65, 94, 101]. There are many ways to represent LPTV systems: common systems include difference equations or state-space models with periodically time-varying coefficients, equivalent to LTI models, obtained by the blocking technique. For an LTI system, as well as for an LPTV one, we can relate the inputs and outputs of the system with a transfer matrix. For an LPTV system, we may use the block model of the system [25, 59, 67], as well.

Chapter 4 presents an iterative AR system parameter estimation algorithm in the presence of white observation noise. The algorithm is based on the parameter estimation bias correction approach. We use high order Yule-Walker equations, sequentially estimate the noise variance, and exploit these estimated variances for the bias correction. The improved performance of the proposed algorithm in the presence of white noise is demonstrated via Monte Carlo experiments. Estimation of the parameters of the AR system from noisy observations of its output has important applications in different fields such as signal processing, speech analysis, spectral estimation, and noise cancellation [51, 97], Kay (1993). However, the estimation of the AR parameters, using noisy data is an important problem in practice [61]. The quality of estimated AR parameters can be severely deteriorated due to the presence of noise. The LS approach will give biased estimates of the AR parameters when the measured signal is corrupted by noise [98]. The techniques of parameter estimation using Yule-Walker equations in the autocorrelation domain are described by [48, 115]. In such a case, parameter estimation is mainly confined by the 5 dB signal to the SNR. The technique requires estimation of noise, which is difficult to obtain, and the autocorrelation function becomes noisy at a low SNR [60, 99]. On the other hand, the use of HO(Y-W) equations does not require a priori estimation of noise, but it is sensitive to the penetration of noise in the positive lag of data samples in the autocorrelation function. These equations do not yield acceptable results under noisy conditions [113]. An LS based method has been developed for noisy AR signals. It does not need to prefilter noisy data [116]. The improved LS method with no prefiltering (ILSNP) yields good results, as compared with the Y-W equations [115]. However, this technique cannot reach 0 dB SNR. An approach for system identification at extremely low SNR, using energy density in the DCT domain, is analyzed by [32]. In the proposed method, the system properties such as the energy density in a DCT domain, have been utilized to estimate the autocorrelation function. Successive autocorrelations of this estimated function are taken for a sequential estimation of system parameters.

Chapter 5 has set the deal on nonlinear systems. The last decade of past century have witnessed a considerable activity in nonlinear modelling [28, 71, 74, 87]. The reason for such an interest is that the most natural systems are nonlinear [18]. New techniques continue to develop and this field expands both in the theoretical and application directions. Nonlinearities have been viewed as deviations from the linear behavior. In many approaches locally-linear or PWL approximations are used to quantify the nonlinear behavior (Fig. 1.1). A variety of nonlinear filtering algorithms has been developed. Unlike linear filters which are completely characterized by the impulse response, it is difficult to find a unified framework for describing nonlinear filters. There is a great variety of nonlinear system structures. The first task is to determine the most appropriate structure for the underlying problem. Good reviews of nonlinear system structures can be found in [9, 63, 91, 92]. Among the nonlinear filters, polynomial filters, i.e., Volterra, polynomial autoregressive, and bilinear filters have been one of the most widely used nonlinear system representation [9]. It is known that the Volterra series expansion has been the most popular model in use for the last four decades. However, one of the problems related polynomial nonlinear system with memory, is the computational complexity. As the polynomial order and/or memory increases, the number of parameters in the polynomial filter also increases rapidly [87]. As a consequence, large data sets required to obtain huge number of model parameter estimates with reasonable accuracy. On the other hand, it is difficult to change the order of nonlinearity.

There are two main approaches for reducing the Volterra filter complexity [87]. The Volterra filter is approximated using a cascade structure, composed of linear filters, in series with memoryless nonlinearities [34, 62]. Another approach is the tensor basis approximation, which represents the Volterra kernel as a linear combination of tensor products of basis vectors [24, 73]. Priestley [75] proposed to represent a nonlinear time series as linear models with variable parameters. He analyzed exponential AR, threshold AR, and bilinear models and showed that they were special cases of the state dependent models.

It is shown in Chap. 5 that nonlinear Volterra, polynomial autoregressive, and bilinear filters have the same layered implementation procedure. Using the layered structure, the order of nonlinearity can be increased by adding more layers to the structure. The structure is modular and consists of the simple MA or AR filters which can be added to the structure to achieve the desired degree of complexity. In addition, the modular layered structures admit the VLSI implementation of polynomial nonlinear filters.

References [11, 12] proposed to identify nonlinear systems as combination of dynamical linear and static nonlinear elements. Nowadays, a lot of physical nonlinear systems are naturally described as block-oriented Hammerstein and Wiener systems (Figs. 1.2, 1.3), which are subdivided into linear dynamic and nonlinear static blocks, respectively. The Wiener system consists of a linear dynamic part that follows a static nonlinearity $f(\cdot, \eta)$, while the Hammerstein system consists of an interconection of a linear dynamic part, precedental by a static nonlinear function. As it is emphasized in [63, 64] block-oriented nonlinear systems appear in many engineering applications. Therefore, the identification of Hammerstein and Wiener systems has been an active research area for many years. Various approaches have been used to identify and to control Hammerstein or Wiener systems: [15, 37, 46, 72, 103, 104, 117]. There is a considerable amount of literature devoted to the identification of nonlinear block-oriented systems. It could be mentioned here just the negligible part of papers: [9, 10, 13, 40, 41, 45, 47, 49, 52, 56, 57, 68, 88, 103–109, 112]. Moreover, a lot of techniques have been worked out to compensate actuator nonlinearities, such as, a friction, deadzone, saturation, preload, backlash, and hysteresis [2, 4, 5, 17, 23, 27, 58, 77–84, 89, 90]. They have been tried to adjust the performance of control systems by reducing non-desired effects of a hard nonlinearity.

In practice, the matter of great interest is in approaches based on the use of self-tuning controllers, represented by [1], that are computationally efficient, speedy, and easily implemented, because the calculation time for digital controllers is usually between a hundredth of a second and one second [14]. Optimization based adaptive controllers are closely linked with the input-output representation of the LTI system. They are coupled with a generalized output error variance to be minimized. It is known that adaptive controllers can be effective in real situations [1, 3, 21, 42–44,

111] (Clarke and Gawthrop 1978). On the other hand, a neural network and a fuzzy logic models are broadly used for self-tuning control of nonlinear processes (see, [89, 90]), too. However, such model representations are frequently too complex and difficult to implement. Moreover, their use to adapt the changes in real processes is problematic, as well [71]).

Among the nonlinear systems there appear, so named, switched PWA systems that are able to change their working regimes (Fig. 1.1) [6, 16, 30]. The switched system consists of several affine subsystems, between which switchings occur at different time moments occasionally. Switched systems are created by both continuous and discrete components. It is known that the switched systems appear in many applications, for example, when there are physical limits to the functioning system or bounds to the control signal, or deadzones, etc. [88]. Chemical or nuclear reactors, and partly or fully-autonomous remote-controlled robots can be examples of switched systems. Unmanned remote working vehicles that are linked to automated steering functions under circumstances could be the following example [88].

The basic idea behind PWL regression used to represent the data of system by mathematical expressions could be presented as follows: 'if the data follow different linear trends over different regions of the data, then we should model regression function in pieces'—see online/stat.psu.edu/stat501/lesson/8/8.10 [120] as well as in papers written by [77, 102]. This method allows us to use an ordinary linear controller for each sector obtained from this approximation according to [8, 36].

Note that any sufficiently smooth nonlinear function can be arbitrarily approximated by the PWA function. A very simple example of the PWA system could be a linear system, controlled by a linear feedback, only where the control signal is bounded [88]. Continuous time PWA system could be described by a state-space mathematical model as folows

$$\dot{\mathbf{x}} = \mathbf{A}(\nu)\mathbf{x} + \mathbf{B}(\nu)\mathbf{u} + \mathbf{b}(\nu),$$

$$\mathbf{y} = \mathbf{C}(\nu)\mathbf{x} + \mathbf{D}(\nu)\mathbf{u} + \mathbf{d}(\nu), \tag{1.1}$$

where $\mathbf{x}$ is the state vector, $\dot{\mathbf{x}}$ is its derivative, $\mathbf{A}$, $\mathbf{B}$, $\mathbf{C}$, $\mathbf{D}$ are respective matrices, $\mathbf{u}$ is the input vector, and $\mathbf{y}$ is the system output, $\mathbf{b}(\nu)$, $\mathbf{d}(\nu)$ are additive noise vectors. Vector ν describes in what mode (i.e., in what affine subsystem) the ARMA system is for the current time t. The vector ν can be a function of $\mathbf{x}$, t, $\mathbf{u}$ or some other external input. It takes a finite number of values. A common restriction is that ν depends only on $\mathbf{x}$.

The PWL function, characteristics of which is shown in Fig. 1.1, also belongs to a certain type of the class of PWA functions. In Fig. 1.4, two estimate types of the hinge PWL function are shown. The first type is based on the ordinary dependence of output $y(k)$ on the input $u(k)$ (Fig. 1.4a, c). The second type is based on the same dependence, using the same signals, the values of which are reordered in an ascending order (Fig. 1.4b, d). In Fig. 1.4 `sort` is the Matlab routine. Graphs, presented in Fig. 1.4, confirm that the second type estimate (parts b, d) is preferable. The forms and breakpoints of PWL hinge functions (b, d) are clearly visible as compared to

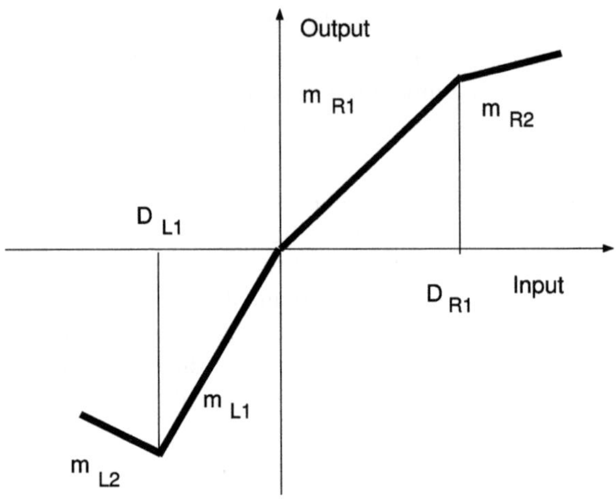

Fig. 1.1 The PWL function having four segments. m_{L1}, m_{L2} m_{R1}, m_{R2} are the corresponding segment slopes, while D_{R1}, D_{L1} are respective constants

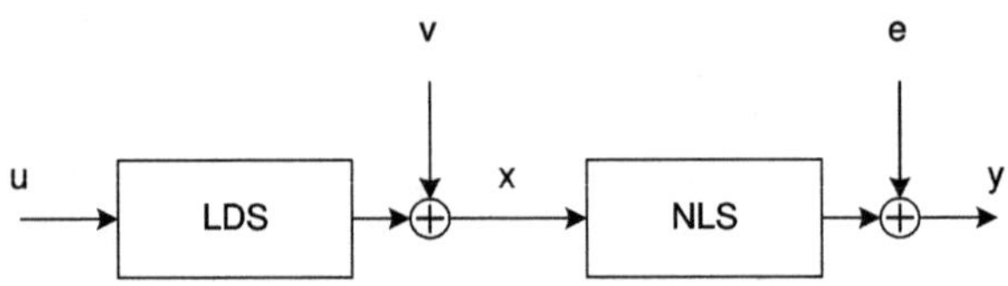

Fig. 1.2 The Hammerstein system with the process noise $w(k)$ and that of the measurement noise $v(k)$. The linear dynamic part $G(q^{-1}, \Theta)$ of the Hammerstein system is parametrised by Θ, while the static nonlinear part $f(\cdot, \eta_0)$—by η_0. Signals: $u(k)$ is input, $y(k)$ is output, $x(k)$ is an unmeasurable intermediate signal; q^{-1} is a backward shift operator

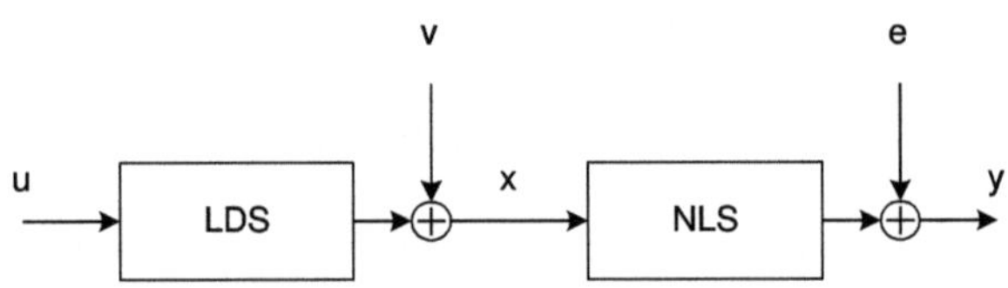

Fig. 1.3 The Wiener system. The values and notation are the same as in Fig. 1.2. The inverse location of linear and nonlinear parts is the main structural difference of the Wiener system in respect of the Hammerstein one. Signals $x(k)$ in both systems are not the same

functions, given in (a, c). Functions (b, d) can be partitioned into three linear segments (Fig. 1.4b) easily. Afterwards, the thresholds, discriminating several parts of PWL hinge functions, can be found. Thus, a simple way to obtain linear system with PWL nonlinearity breakpoint values might be the one that determines the dependence of the measured output $y(n)$ on the input $x(n)$, see Fig. 1.4. After polyhedral partition of the state-space of third order nonlinear system into distinct hinging hyperplane functions, we have one or several possible affine systems in each region (Fig. 1.5).

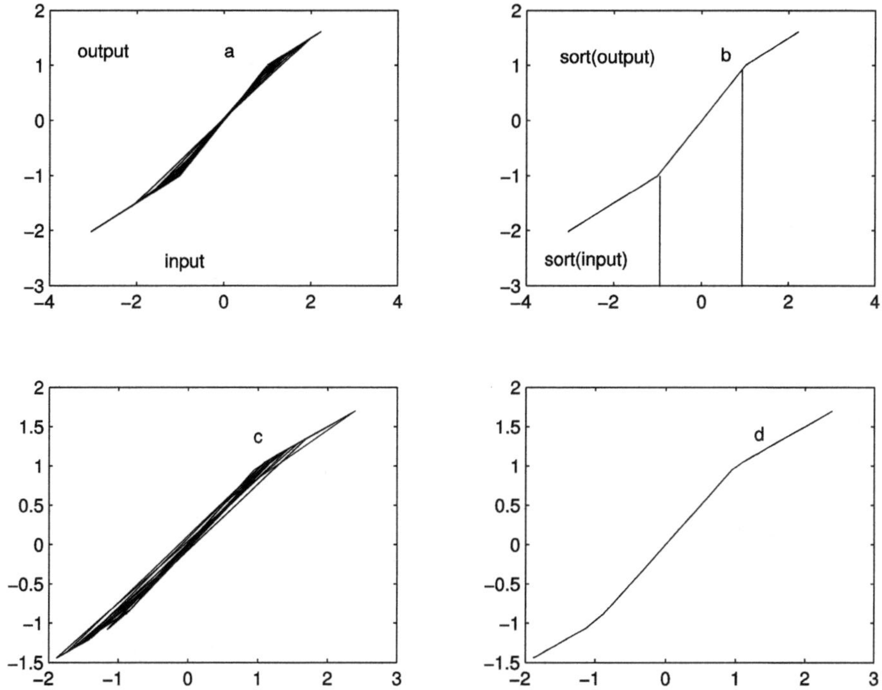

Fig. 1.4 Estimates of the three-segment hinge function: output $y(k)$ in dependence on the input $x(k)$ (**a**, **c**); $\mathtt{sort}(y(k))$ in dependence on the input $\mathtt{sort}(x(k))$ (**b**, **d**)

Switchings between available states can occure if a border of a polyhedral region is reached [88].

Let us assume that the continuous affined subsystems, the dynamics of which is described by Eq. (1.1) are located inside the plant. Subsystems are related by means of the generator [88] (see Fig. 1.6), which contains information about hyperplanes and generates logical input symbols $e^*_{s_i}$ for a discrete controller. The actuator translates controller's output symbols ν^* into the key vector signal ν, which describes in what mode, i.e., in what affined subsystem position, the initial system (plant) will be at the present time [88]. If the value of the signal ν changes, the system switches to another mode, i.e., to another affined subsystem.

The general discrete time PWA system could be described as follows [88]

$$\mathbf{x}(n+1) = \mathbf{A}(\nu)\mathbf{x}(n) + \mathbf{B}(\nu)\mathbf{u}(n) + \mathbf{b}(\nu),$$

$$\mathbf{y}(n) = \mathbf{C}(\nu)\mathbf{x}(n) + \mathbf{D}(\nu)\mathbf{u}(n) + \mathbf{d}(\nu). \quad (1.2)$$

Here $n = T_s n$, $\forall\, n = 0, 1, 2, \ldots$ is the discrete time, where T_s is a sampling period to be chosen according to the Nyquist sampling theorem [76], $\mathbf{x}(n)$ is the state vector, $\mathbf{u}(n)$ is the input vector, and $\mathbf{y}(n)$ is the system output. The vector ν can be a function

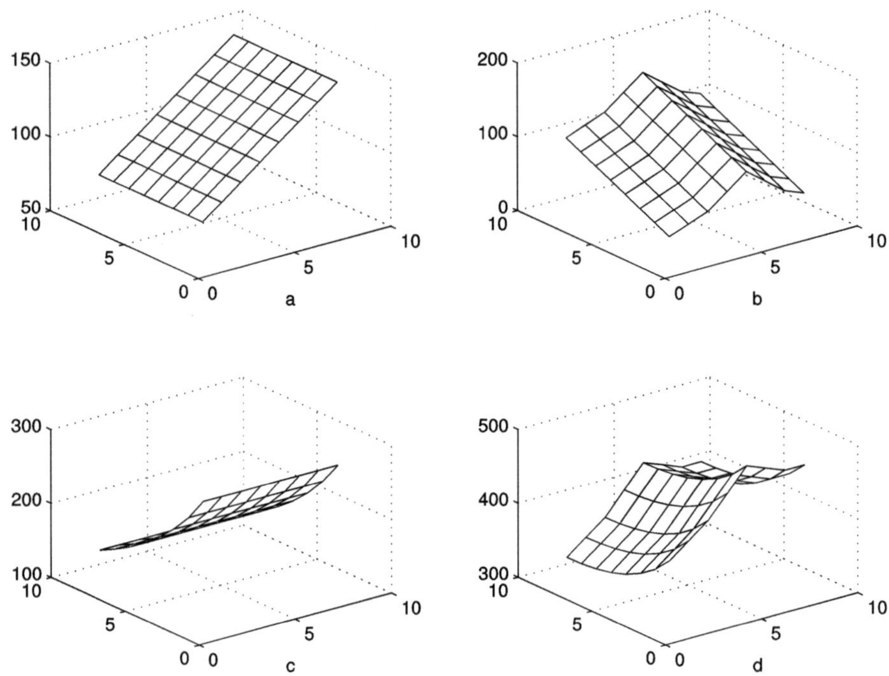

Fig. 1.5 Examples of hinging hyperplanes in (**a, b, c**) of PWL function of third order. The resulting hyperplane of the PWL function is (**d**)

Fig. 1.6 A flowchart of a switched closed-loop system

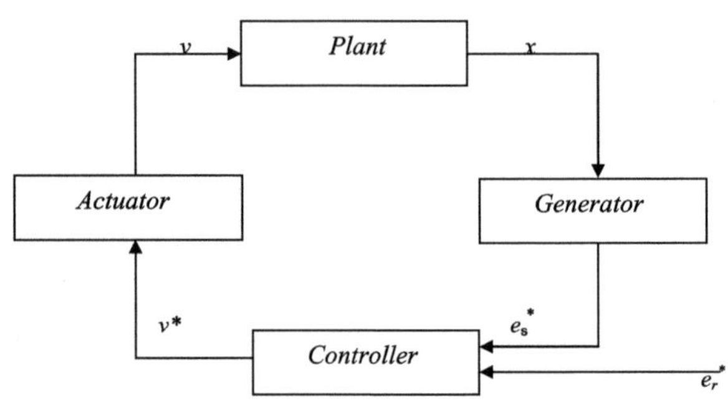

of $\mathbf{x}(n)$, n, $\mathbf{u}(n)$ or some other external input. Common restrictions are that ν takes a finite value number and that it depends only on $\mathbf{x}(n)$.

The universal requirement to the behavior of the closed-loop system is that it should be stable [4, 71]. Unfortunately, the stability issue is most often a complicated matter for nonlinear systems. For example, it may occur that a nonlinear system locally is unstable in certain regimes while it is stable in others ones. Analysis of nonlinear dynamics and 'design of globally stabilizing controllers for nonlinear systems are topics, which are addressed in many textbooks' [29, 71]. The location of poles and zeros of a direct system has important influence to the stability of control process, when linear block of discrete-time nonlinear system is modelled by respective transfer function. The global stability of the controlled system requires that its inverse system should be stable as well. Such requirement excludes non-minimum phase systems [3].

The switched systems ought to meet additional requirement for Wiener systems, as follows: the resulting algorithms require that the static nonlinearity function should be invertible. This requirement restricts the applicability of switched PWL systems strongly enough. Note that it is not needed to have the inverse function of PWL nonlinearity in the parameter estimation flowchart if the initial system is Hammerstein's one. However, it is important to know the structure of PWL nonlinearity in advance, in order to substitute the current parameter estimates instead of unknown true parameter values. We will try here to work out the parametric identification approach, based on the data partition, that should avoid available restriction. Namely, the approach to be proposed here could be used not only for a potential system having the invertible PWL nonlinearity, it ought to be applicable for a system with a noninvertible nonlinear function, as well. Note that abovementioned notation does not concern nonlinear systems having non-minimum phase linear blocks. In this connection, questions that will be explored in Chaps. 6–8 and 9, are as follows:

(a) How efficient is the approach, based on the data partition?
(b) How system and noise filter parameter estimates, determined using the data partition, have influenced the stability of the servo control?
(c) How the accuracy of control of block-oriented system having a PWL nonlinearity is sensitive to the intensity of process and measurement noises?

In Chap. 6 the available potentiality of ARMAX system having PWL nonlinearity is investigated in respect of MV/GMV control accuracy. ARMA and/or ARMAX systems related with PWL nonlinearities have deserved a growing interest as PWA systems, partly, because of their close relationship to hybrid ones. It is assumed that a stochastic system is represented by state-space equations, including the known static invertible nonlinearity. The steady-state innovations model in observer canonical form is expressed in the form of n-dimensional ARMAX model. Then, the problem of parametric estimation of the observer state-space form is solved here using n-dimensional ARMAX model. The LTI system's controller can be determined, as usual, by minimizing the exponentially stable quadratic performance function. Further, the controller expression is recalculated in each current iteration using the linear and nonlinear blocks parameter estimates and input-output samples. Two types of nonlinear block-oriented systems, the Hammerstein system and the Wiener one, are analysed in closed-loop framework. It is shown here that the direct approach could be used for a closed-loop LTI system to estimate ARMA model parameters as well as for the nonlinear Wiener system if respective inverse filter is known. The use of abovementioned filter in the parameter estimation flowchart is extremely important. However, the inverse filter will have negative influence on the accuracy of estimates if the parametric estimation algorithm is working in a noisy environment. The various simulation results of servo control of Wiener system are obtained under following circumstances: the nonlinearity is absent; the nonlinearity is present but it is not registered; the nonlinearity is present and it is known. Parameters of the present nonlinearity as well as that of the linear block are unknown and ought to be determined. In the last experiment the MV controller was changed by the GMV one, fixing their influence on the servo control quality in respect of conformity of system output signal to the reference signal.

In Chap. 7, a parametric identification problem of a class of nonlinear block-oriented systems, consisting of a linear block and a static PWL nonlinearity is solved, using input-output observation samples. The problem is not trivial because a system's output has mixed samples of three subsystems, between which occasional switchings happend at different time moments. To identify a nonlinear system parametrically it is necessary to remix samples in a true order. Then, Wiener system output $y(k)$ observation samples could be partitioned into distinct data sets according to different descriptions. However, the boundaries of observation sets of $y(k)$ depend on the value of unknown thresholds. If thresholds are not known, it is important to determine their estimates. Then, observations can be divided according to distinct regimes, subject to whether some threshold variable determined is smaller or larger than the true threshold [77, 88, 106–109]. Therefore, the problem of identification of unknown parameters of Wiener and/or Hammerstein systems could be solved, if a simple way of determining the thresholds is known. Such a way to obtain threshold values for the Wiener or Hammerstein system is to determine a plot of input $x(n)$ against the measured output $y(n)$, as it is shown by graphs in Fig. 1.4. After right partitioning the data the nonlinear parametric identification problem turns out to the ordinary multifold linear problem [77]. The FIR regression models number corresponds to the PWL nonlinearity segments number. Simulation results confirm the ability of the ordinary LS algorithms to find parameters of multifold regression, generating distinct sets of FIR model coefficients for each segment available.

Chapter 8 is allocated to the development of an approach for parametric identification of Wiener and Hammerstein systems with deadzone nonlinearity. It is shown here that by a simple data reordering and by the following data partition the problem of identification of nonlinear Wiener and Hammerstein systems could be reduced to a linear parametric estimation problem, too. Afterwards, the estimates of unknown parameters of linear regression models are calculated by processing the respective particles of input-output data. A technique based on ordinary PLR method, was proposed here for the estimation of parameters of linear and nonlinear parts of Wiener and Hammerstein systems, including the unknown threshold of PWL characteristics, as well. The results of numerical simulation and identification, obtained by processing observation samples of input-output signals of a discrete-time Wiener and Hammerstein systems with deadzone nonlinearities are given by computer. In order to determine how realizations of different process and measurement noises affect the accuracy of estimation of unknown parameters, we have used the Monte Carlo simulation with 10 data sets, each containing 100 pairs of input-output observations. 10 experiments with the same realization of the process noise and different realizations of the measurement noise with different levels of their intensity have been carried out. The intensity of noises was assured by choosing respective SNR.

As training signals for given nonlinearities, the periodical signal and white Gaussian noise with variance unity were chosen. In each i-th experiment the estimates of parameters were calculated. During the Monte Carlo simulation the averaged estimate values of parameters and of the threshold with their confidence intervals were determined.

Chapter 9 has set a great deal of MVC of Wiener systems, operating in a noisy environment. Results of numerical simulation of MV STC for the Wiener system having invertible as well as noninvertible PWL nonlinearities are represented. Simulation results are given for Wiener system in an unnoisy and noisy frames. Two different types of MV control could be used for nonlinear systems with distinct PWL nonlinearities: adaptive control in a case of an invertible nonlinearity and non-adaptive one for both types of nonlinearities. During the adaptive control recursive linear block parameter PLR estimates are used to calculate the controller parameter values sequentially. The processing time of both algorithms ought to fit into the same sampling period T_s. It should be chosen carefully because it is known [3] that zeros outside the unit circle can be awaked while the sampling of linear continuous-time equations. It is to be expected [71] that a similar behavior can be found in discrete models of nonlinear systems. The non-adaptive control type is used in the absence of PWL nonlinearity inverse function as well. It realizes one-shot controller tuning by the use of stored in computer's memory reordered data, accordingly each segment of PWL nonlinearity. Recursive parametric identification and controller parameter calculation routines are connected only in off-line regime during the training time. Once the training time is over, the parametric identification is stopped until the beginning of next connection. Parameter vectors $\tilde{\theta}(\nu_j)$ can be determined numerically, using two methods: an off-line method and on-line one. Both methods, acordingly to [71], can be named, as follows, first method, as general training one and second method, as specialized training method. In both cases PLR method could be efficient: batch PLR method while the general training, and recursive PLR technique during the specialized training. The recursive approach, proposed in Pupeikis, Kazlauskas 2021, allows us to avoid repeating arithmetical operations in each current iteration. The algorithm presented here, could be effective in real-time applications for very large N and the relatively small kernel $h(n)$. The linear convolution, after a slight modification, turns out to an efficient adaptive procedure that reduces arithmetical operations, sufficiently. Further, it could be accelerated by resolving the main procedure into elementary adaptive algorithms, working in parallel.

References

1. Agarwal, M., Seborg, D.E.: Self-tuning controllers for nonlinear systems. Automatica **1987**, 209–214 (1987)
2. Armstrong-Hélouvry, B., Dupont, P., Canudas de Wit, C.: A survey of models, analysis tools and compensation methods for the control of machines with friction. Automatica **30**(7), 1083–1138 (1994)

3. Aström, K.J., Hagander, P., Sternby, J.: Zeros of sampled system. Automatica **20**(1), 31–38 (1984)
4. Aström, K.J., Wittenmark, B.: Computer-Controlled Systems: Theory and Design. Prentice Hall, NJ (1984)
5. Bai, E.W.: Identification of linear system with hard input nonlinearities of known structure. Automatica **38**(2002), 853–860 (2002)
6. Bemporad, A., Morari, M.: Control of systems integrating logic, dynamics, and constraints. Automatica **35**(3), 407–427 (1999)
7. Berger, I.O., Wolpert, R.: The Likelihood Principle. Harward, CA, Inst Math Stat (1988)
8. Biagola, S., Agamennoni, O., Figueroa, J.: H_∞ control of a Wiener-type system. Int. J. Control **77**(6), 57–583 (2004)
9. Billings, S.A.: Identification of nonlinear systems - a survey. Proc Inst Elect Eng **127**, 272–285 (1980)
10. Billings, S.A., Fakhouri, S.Y.: Identification of nonlinear systems using the Wiener model. Electron. Lett. **13**(17), 502–504 (1977)
11. Billings, S.A., Fakhouri, S.Y.: The theory separable processes with application to the identification of nonlinear systems. ACSE Research Report no 65, University of Sheffield (1978a). eprints.whiterose.ac.uk/87083/
12. Billings, S.A., Fakhouri, S.Y.: Nonlinear system identification using the Hammerstein model. ACSE Research Report no 68, University of Sheffield (1978b). eprints.whiterose.ac.uk/75422/
13. Billings, S.A., Fakhouri, S.Y.: Identification of systems containing linear dynamic and static nonlinear elements. Automatica **18**(1) (1982)
14. Bobál, V., Böhm, J., Fessl, J., Machaček, J.: Digital Self-Tuning Controllers. Springer, Berlin (2005)
15. Bloemen, H.H.J., Chou, C.T., Van den Boom, T.J.J., Verdult, V., Verhaegen, M., Backx, T.C.: Wiener model identification and predictive control for dual composition control of a distillation column. J. Proc. Control **11**(6), 601–620 (2001)
16. Branicky, M.S.: Multiple Lyapunov functions and other analysis tools for switched and hybrid systems. IEEE Trans. AC **43**(4), 475–482 (1998)
17. Canudas de Wit, C., Olsson, H., et al.: A new model for control of systems with friction. IEEE Trans. AC **40**(3), 419–425 (1995)
18. Celka, P., Bershad, N.J., Vesin, J.: Stochastic gradient identification of polynomial Wiener systems. IEEE Trans. Sign Proc. **49**(2), 301–313 (2001)
19. Clark, G.A., Mitra, S.K., Parker, S.R.: Block implementation of adaptive digital filters. IEEE Trans. Circuits Syst. **28**, 584–592 (1981)
20. Clark, G.A., Mitra, S.K., Parker, S.R.: A unified approach to time and frequency domain realization of FIR adaptive digital filters. IEEE Trans. ASSP **31**, 1073–1083 (1983)
21. Clarke, D.W., Gawthrop, P.J.: Self-tuning controller. Proc. IEEE (122), 929–934 (1975). https://doi.org/10.1049/piee.1975.0252
22. Clarke, D.W., Gawthrop, P.J.: Self-tuning control. Proc. IEEE **126**(6), 929–934 (1979). https://doi.org/10.1049/piee.1979.0145
23. Chan, C.W., Hui, K.: On the existence of globally stable actuator saturation compensators. Int. J. Control **69**(6), 773–788 (1998)
24. Chiang, H., Nikias, C.L., Venetsanopoulos, A.N.: Efficient implementation of quadratic digital filters. IEEE Trans. ASSP **34**(6), 1511–1528 (1986)
25. Chen, T., Qiu, L.: Linear periodically time-varying discrete-time systems: Aliasing and LTI approximations. Syst. Control Lett. **30**(5), 225–235 (1997)
26. Dentino, M., McCool, J., Widrow, B.: Adaptive filtering in the frequency domain. Proc. IEEE **66**(12), 1658–1659 (1978)
27. Desoer, C.A., Shahruz, S.M.: Stability of dithered nonlinear systems with backlash or hysteresis. Int. J. Control **43**(4), 1045–1060 (1986)
28. Dokic, M.V., Clarkson, P.M.: On the performance of a second-order adaptive Volterra filter. IEEE Trans. Signal Process. **41**(2), 1944–1947 (1993)

29. Dhillon, A., Verma, G.K.: Convolutional neural network: a review of models, methodologies and applications to object detection. Progress Artif. Intell. **9**(2020), 85–112 (2020)
30. Einarsson, V.: Model checking methods for mode switching systems. Ph.D. thesis, Department of Electrical Engineering, Linköping University, Linköping Sweden (2000). https://www.control.isy.liu.se
31. Eykhoff, P.: System Identification, Parameter and State Estimation. Wiley, NJ (1974)
32. Ferdousi, F., Connie, A.T., Sharmin, M., Khan, M.H.: System identification at an extremely low SNR using energy density in DCT domain. IEEE Signal Process. Lett. **12**(4), 289–292 (2005)
33. Ferrara, E.R.: Fast implementation of LMS adaptive filters. IEEE Trans. ASSP **28**, 474–475 (1980)
34. Frank, W.: MMD-A_n efficient approximation to the second-order Volterra filter. Proc. ICASSP **3**, 517–520 (1994)
35. Gao, W., Selmic, R.R.: Neural network control of a class of nonlinear systems with saturation. IEEE Trans. NN **17**(1), 147–156 (2006)
36. García, A.G., Agamennoni, O., Figueroa, J.: Applying continuous piecewise linear approximations to affine nonlinear control systems. IFAC Proc. **42**(6), 114–119 (2009)
37. Giri, F., Bai, E.W. (eds.).: Block-oriented nonlinear system identification. In: Lecture Notes in Control and Information Sciences. Springer, Berlin (2010). https://doi.org/10.1007/978-1-849969-513-2
38. Goodwin, G.C., Sang Sin, K.: Adaptive Filtering, Prediction and Control. Prentice Hall, NJ (1984)
39. Gonzales, R.C., Woods, R.E.: Digital Image Processing, 2nd edn. Prentice-Hall, NJ (2002)
40. Greblicki, W., Pawlak, M.: Hammerstein system identification by non-parametric regression estimation. Int. J. Control **45**(1), 343–354 (1987)
41. Greblicki, W., Pawlak, M.: Nonparametric identification of nonlinear block-oriented systems. In: Proceedings of of 9th IFAC/IFORS Symposium on Identification and System Parameter Estimation, pp. 720–724. Budapest (1991)
42. Grimble, M.J.: Implicit and explicit LQG self-tuning controllers. Automatica **20**(5), 661–669 (1984)
43. Grimble, M.J., Zaknich, A., Johnson, M.A. (eds.).: Principles of Adaptive Filters and Self-Learning Systems. Springer, Berlin (2005a)
44. Grimble, M.J., Martin, P., Greenwood, D., Moradi, M.H., Johnson, M.A.: PID Control: New Identification and Design Methods. Springer, Berlin (2005)
45. Guo, F., Bretthauer, G.: Identification of MISO Wiener and Hammerstein systems. In: Proceedings of the 7th European Control Conference (TEE'03), pp. 2144–2149. University of Cambrige (2003)
46. Haber, R., Keviczky, L.: Nonlinear System Identification—Input-Output Modelling Approach. Kluwer Academic Publishers, London (1999)
47. Hagenblad, A.: Aspects of the identification of Wieners models. Licentiate thesis No. 793. Department of Electrical Engineering, Linköping University, Linköping Sweden (1999). https://www.control.isy.liu.se
48. Hasan, M.K., Khan, M.R.: Identification of autoregressive systems at a very low SNR using cosine model based estimation of autocorrelation function. J. Signal Process. **7**(1), 15–21 (2003)
49. Hasiewicz, Z., Mzyk, G.: Combined parametric-nonparametric identification of Hammerstein systems. IEEE Trans. AC **49**(8), 1370–1375 (2004)
50. Haykin, S.: Adaptive Filter Theory. Prentice-Hall, Englewood Cliffs, NJ (1986)
51. Haykin, S.: Adaptive Filter Theory. Prentice-Hall, Englewood Cliffs, NJ (1996)
52. Hunter, I.W., Korenberg, M.J.: The identification of nonlinear biological systems: Wiener kernel approaches. Ann. Biomed. Eng. **18**, 629–654 (1990)
53. Isermann, R.: Practical aspects of process identification. Automatica **16**(5), 575–587 (1980)
54. Isermann, R.: Digital Control Systems. Springer, Berlin (1981)
55. Jackson, L.B.: Digital Filters and Signal Processing. Kluwer Academic (1986)

56. Janczak, A.: Parameter estimation based fault detection and isolation in Wiener and Hammerstein systems. Int. J. Appl. Math. Comput. Sci. **9**(3), 711–735 (1999)
57. Kalafatis, A.D., Wang, L., Cluett, W.R.: Principles and Application. Kluwer Academic (1997)
58. Kapila, V., Grigoriadis, K.M.: Actuator Saturation Control. Marcel Dekker Inc, NY (2002)
59. Kazlauskas, K.: Pipelined-block models of linear discrete-time linear periodically time-varying systems. Informatica, LT MII VU Press Vilnius **8**(2), 215–236 (1997)
60. Kay, S.M.: Noise compensation for autoregressive spectral estimates. IEEE Trans. ASSP **28**(3), 292–303 (1980)
61. Kay, S.M.: Modern Spectral Estimation. Englewood Cliffs, Prentice-Hall, NJ (1988)
62. Korenberg, M.J.: Orthogonal approaches to time-series analysis and system identification. IEEE Signal Process. Mag. **8**(1), 29–43 (1991)
63. Ljung, L.: System Identification. Theory for the Users. Englewood Cliffs, Prentice-Hall, NJ (1987)
64. Ljung, L.: System Identification. Theory for the Users, 2nd edn. Prentice-Hall (1999)
65. Mehr, A.S., Chen, T.: On alias-component matrices of discrete-time linear periodically time-varying systems. Signal Process. Lett. **8**(4), 114–116 (2001)
66. Meng, T.H., Messerschmitt, D.G.: IEEE Trans. ASSP **34**(4), 455–470 (1987)
67. Meyer, R.A., Burrus, C.S.: A unified analysis of multirate and periodically time-varying digital filters. IEEE Trans. Circuits Syst. **22**(3), 162–168 (1975)
68. Mielcarek, P., Mzyk, G.: Recursive algorithm for interaction prediction in Hammerstein system identification with experimental sudies. Int. J. Model. Ident. Control **43**(2), 134–144 (2023)
69. Mikhael, W.B., Wu, F.H.: Fast algorithms for block FIR adaptive digital filtering. IEEE Trans. Circuits Syst. **34**, 1152–1160 (1987)
70. Mikhael, W.B., Wu, F.H.: A fast block FIR adaptive digital filtering algorithm with individual adaptation of parameters. IEEE Trans. Circuits Syst. **36**, 1–10 (1989)
71. Nørgaard, et al.: Neural networks for modelling and control of dynamic systems. A practitioner's handbook. In: Advanced Textbooks in Control and Signal Processing. Springer (2000)
72. Norquay, S.J., Palazoglu, A., Romagholi, J.A.: Model predictive control based on Wiener models. Chem. Eng. Sci. **53**(1998), 75–84 (1998)
73. Panicker, T.M., Mathews, V.J.: Parallel-cascade realizations and approximations of truncated Volterra systems. In: Proceedings of ICASSP, pp. 412–415 (1996)
74. Pitas, I., Venetsanopoulos.: Nonlinear Digital Filters: Principles and Applications. Kluwer Academic (1990)
75. Priestley, M.B.: Nonlinear and Non-Stationary Time Series Analysis. Academic (1980)
76. Proakis, J.G., Manolakis, D.G.: Digital Signal Processing: Principles, Algorithms and Applications. Prentice Hall, London (1989)
77. Pupeikis, R.: Identification of piecewise affine Wiener systems using data partition. Technical Report LiTH-ISY-R-2554 Depart Electr Engin Linköping Univ SE-581 83 Linköping Sweden (2003)
78. Pupeikis, R.: On the identification of Wiener systems having saturation-like functions with positive slopes. Informatica, LT MII VU Press Vilnius **16**(1), 131–144 (2005)
79. Pupeikis, R.: On the identification of Hammerstein-Wiener systems. Liet mat rink, LT MII VU Press Vilnius **45**(spec nr), 509–514 (2005b)
80. Pupeikis, R.: On the identification of Hammerstein systems having saturation-like functions with positive slopes. Informatica, LT MII VU Press Vilnius **17**(1), 55–68 (2006)
81. Pupeikis, R.: On recursive parametric identification of Wiener systems. ITC LT KTU Press Kaunas **40**(1), 21–28 (2011)
82. Pupeikis, R.: Self-tuning minimum variance control of linear systems followed by saturation nonlinearities in a noisy frame. Int. J. Robust Nonlinear Control **24**(2), 313–325 (2014)
83. Pupeikis, R.: Parametric identification of linear system followed by noninvertible nonlinearities. Int. Front. Sci. Lett. **6**, 16–27 (2015). https://doi.org/10.18052/www.scipress.com/IFSL.6.16

84. Pupeikis, R.: Reinforced Methods for Dynamical System Identification and Adaptive Control. Shaker Verlag GmbH, Aachen (2017)
85. Pupeikis, R., Navakauskas, D., Ljung, L.: Identification of Wiener systems with hard and discontinuous nonlinearities. Technical Report LiTH-ISY-R-2501 Department of Electrical Engineering Linköping University SE-581 83 Linköping, Sweden (2003)
86. Pupeikis, R., Kazlauskas, K.: Advanced Methods for Digital Systems and Signal Processing. Shaker Verlag GmbH, Düren (2021)
87. Raz, G.M., Van Veen, B.D.: Baseband Volterra filters for implementing carier based nonlinearities. IEEE Trans. Signal Process. **6**(1), 103–114 (1998)
88. Roll, J.: Local and piecewise affine approaches to system identification. Linköping stud sci techn. Diss No 802. Depart Electr Engin Linköping Univ SE-581 83 Linköping, Sweden (2003). https://www.control.isy.liu.se
89. Selmic, R.R., Lewis, F.L.: Neural net backlash compensation with Hebbian tuning using dynamic inversion. Automatica **37**(2001), 1269–1277 (2001)
90. Selmic, R.R., Phoha, V.V., Lewis, F.L.: Intelligent compensation of actuator nonlinearities. In: Proceedings of 42nd IEEE Conference on Decision and Control, pp. 4327–4332. Maui (2003). https://www2.latech.edu/~phoha/PDF
91. Schetzen, M.: The Volterra and Wiener Theories of Nonlinear Systems. Malabar, FL Krieger (1980)
92. Schoukens, J., Ljung, L.: Nonlinear System Identification. A user-oriented road map. Prepr submitted to IEEE Control Magazine. Cornel University (2019). arxiv:1902.00683
93. Schoukens, J., Pintelon, R., Dobrowiecki, T., Rolain, Y.: Identification of linear systems with nonlinear distortions. Automatica **41**(2005), 491–504 (2005)
94. Shenoy, R.G., Burnside, D., Parks, T.W.: Linear periodic systems and multirate filter design. IEEE Trans. Signal Process. **42**(9), 2242–2256 (1994)
95. Sledevič, T., Tamulevičius, G., Navakauskas, D.: Upgrating FPGA implementation of isolated word recognition system for a real-time operation. Electron Electr. Eng. **19**(10), 123–128 (2013)
96. Sledevič, T., Serackis, A., Tamulevičius, G., Navakauskas, D.: Evaluation of feature extraction algorithms for a real-time isolated word recognition system. Int. J. Electron. Sci. Eng. **7**(12), 303–307 (2013)
97. Srinath, M.D., Rajasekaran, P.K., Viswanathan, R.: Introduction to Statistical Signal Processing with Applications. Prentice-Hall Englewood Cliffs, NJ (1996)
98. Stoica, P., Nehorai, A., Kay, S.M.: Statistical analysis of the least squares autoregressive estimator in the presence of noise. IEEE Trans. ASSP **35**(9), 1273–1281 (1987)
99. Sun, W., Yahagi, T.: A new over fitting lattice filter for ARMA parameter estimation and additive noise. IEICE Trans. Fund. **E75-A**(2), 247–254 (1992)
100. Tham, M.T.: Minimum Variance and Generalized Minimum Variance Control Algorithms. University of Newcastle upon Tyne (1999). https://lorien.ncl.ac.uk/ming/digicont/mbpc/gmv.pdf
101. Vaidyanathan, P.P.: Multirate Systems and Filter Banks. Englewood Cliffs, Prentice-Hall, NJ (1993)
102. Van Pelt, T.H., Bernstein, D.S.: Nonlinear system identification using Hammerstein and nonlinear feedback models with piecewise linear static maps-part 1:theory. In: Proceedings of American Control Conference, pp. 225–229 (2000)
103. Verhaegen, M., Westwick, D.: Identifying MIMO Hammerstein systems in the context of subspace model identification methods. Int. J. Control **63**(2), 331–349 (1996)
104. Verhaegen, M., Westwick, D.: Identifying MIMO Wiener systems using subspace model identification methods. Signal Process. **52**(2), 235–258 (1996)
105. Vörös, J.: Parameter identification of discontinuous Hammerstein systems. Automatica **33**(6), 1141–1146 (1997)
106. Vörös, J.: Parameter identification of Hammerstein systems with discontinuous nonlinearities. Syst. Control Lett. **44**, 364–372 (2001)

107. Vörös, J.: Modelling and parameter identification of systems with multisegment piecewise-linear characteristics. IEEE Trans. Autom. Control **47**(1), 184–188 (2002)
108. Vörös, J.: Recursive identification of systems with noninvertible output nonlinearities. Informatica LT MII VU Press **21**(1), 139–148 (2010)
109. Vörös, J.: Block-oriented nonlinear system identification. In: Lecture Notes in Control and Information Sciences, vol. 404, pp. 35–51 (2010b)
110. Wang, T., Wang, C.L.: On the optimum design of the block adaptive FIR digital filter. IEEE Trans. Signal Process. **41**, 2131–2140 (1993)
111. Warwic, K., Irvin, G.W., Hunt, K.J.: Neural network for control and systems. IEEE Workshop IEEE Control Eng. Ser. **31**(1), 260 (1992)
112. Wigren, T.: Recursive prediction error identification using the nonlinear Wiener model. Automatica **29**(4), 1011–1025 (1993)
113. Yahagi, T., Hasan, M.K.: Estimation of noise variance from noisy measurements of AR and ARMA systems: Application to blind identification of linear time-invariant systems. IEICE Trans. Fund. E **77**(5), 847–855 (1994)
114. Zhang, X.D.: Modern Signal Processing. In the series De Gruyter STEM (2022). https://doi.org/10.1515/9783110475562
115. Zheng, W.X.: A least-squares based method for autoregressive signals in the presence of noise. IEEE Trans. Circuits Syst. II **46**(1), 81–85 (1999)
116. Zheng, W.X.: Autoregressive parameter estimation from noisy data. IEEE Trans. Circuits Syst. **47**(1), 71–75 (2000)
117. Zhu, Y.: Parametric Wiener model identification for control. In : Proceedings of 14th IFAC World Congress H, pp. 37–42. Beijing (1999)
118. Joint special issue on adaptive signal processing. IEEE Trans. ASSP **29** (1981)
119. Special Issue on Adaptive Systems and Applications. IEEE Trans. Circuits Syst. **34** (1987)
120. online/stat.psu.edu/stat501/lesson/8/8.10. Accessed 15 Oct 2022

Chapter 2
High Speed Least Mean Square Adaptive Filtering

In this chapter we show that the LMS algorithm can be speeded up without changing any of its adaptive characteristics. The parallel LMS adaptive filtering algorithm and its modifications are presented. High speed is achieved by increasing the parallelism in the LMS adaptive algorithm through a proper modification of the LMS adaptive algorithm. An iterative procedures for efficient computation of the lower triangular inverse matrix and the input signal covariance matrix are presented.

2.1 The LMS Adaptive Filter

The LMS adaptive filter [3] is an FIR digital filter of order $L - 1$ for which the output $\tilde{y}(k)$ at discrete time instants k is given as the convolution sum of the input $x(k)$ and the filter weights $w_i(k)$:

$$\tilde{y}(k) = \sum_{i=0}^{L-1} x(k-i)w_i(k) = \mathbf{X}^T(k)\mathbf{W}(k) = \mathbf{W}^T(k)\mathbf{X}(k), \quad k = 0, 1, ... \tag{2.1}$$

The LMS algorithm adjusts the filter weights in accordance with (2.2):

$$\mathbf{W}(k+1) = \mathbf{W}(k) + \mu e(k)\mathbf{X}(k), \tag{2.2}$$

where μ is a parameter that controls rate and stability of convergence, and $\mathbf{W}(k)$ and $\mathbf{X}(k)$ are, respectively, the $L \times 1$ weight vector and the $L \times 1$ input vector:

$$\mathbf{W}(k) = [w_0(k), w_1(k), ..., w_{L-1}(k)]^T,$$

$$\mathbf{X}(k) = [x(k), x(k-1), ..., x(k-L+1)]^T, \tag{2.3}$$

R. Pupeikis and K. Kazlauskas, *Advanced Digital Signal Processing Methods for Filtering, Identification, and Nonlinear Systems Control*, Studies in Systems, Decision and Control 638, https://doi.org/10.1007/978-3-032-06740-1_2

and $e(k)$ is the error at the kth instant given by the difference between the desired output $y(k)$ and the actual output $\tilde{y}(k)$:

$$e(k) = y(k) - \mathbf{X}^T(k)\mathbf{W}(k). \tag{2.4}$$

Letting $k = mM + i, i = 0, 1, ..., M - 1$; denote the block number, and $M = 1, 2, ...$ the block length, there are two different vectors of interest for the block adaptive filter. The first is the $L \times 1$ vector $\mathbf{X}(k)$ of input stored in the filter's register at time k from (2.3). The other is the mth input data block designed by the $M \times 1$ vector

$$\tilde{\mathbf{x}}(m) = [x(mM), x(mM + 1), ..., x(mM + M - 1)]^T. \tag{2.5}$$

Similarly, the mth output data block is given by the $M \times 1$ vector

$$\bar{\hat{y}} = [\hat{y}(mM), \hat{y}(mM + 1), .., \hat{y}(mM + M - 1)]^T,$$

and desired mth output data block is given by the $M \times 1$ vector

$$\bar{y} = [y(mM), y(mM + 1), .., y(mM + M - 1)]^T.$$

Since the filter weights are adjusted on a block by block basis, the weights for the mth block are denoted by the following element vector

$$\mathbf{W}(m) = [w_0(m), w_1(m), ..., w_{L-1}(m)]^T, \tag{2.6}$$

where the subscripts denote the individual filter weights.

The block adaptive filter output is given by the equation

$$\bar{\mathbf{y}} = \bar{\mathbf{X}}(m)\mathbf{W}(m), \tag{2.7}$$

where $\bar{\mathbf{X}}(m)$ is an $M \times L$ given by

$$\bar{\mathbf{X}}(m) = \begin{bmatrix} x(mM) & x(mM - 1) & ... & x(mM - L + 1) \\ x(mM + 1) & x(mM) & ... & x(mM - L + 2) \\ ... & ... & ... & ... \\ x(mM + M - 1) & x(mM + M - 2) & ... & x(mM + M - L) \end{bmatrix} \tag{2.8}$$

Substituting (2.6) and (2.8) into (2.7) yields

$$\hat{\mathbf{y}}(mM + l) = \mathbf{X}(mM + l)\mathbf{W}(m), \quad l = 0, 1, ..., M - 1, \tag{2.9}$$

where $\mathbf{X}(mM + l)$ is lth row of the matrix $\bar{\mathbf{X}}(m)$.

Equation (2.9) gives the elements of $\bar{\hat{\mathbf{y}}}$, the mth block of the output data, as a linear convolution, which can also be written as

$$\hat{y}(mM + l) = \sum_{i=0}^{L-1} x(mM + l - i)w_i(m), \quad l = 0, 1, ..., M - 1. \tag{2.10}$$

The filter weights are updated as follows [1]

$$\mathbf{W}(m + 1) = \mathbf{W}(m) + \frac{2\mu_\beta}{M}\hat{\mathbf{X}}^T(m)\bar{e}(m), \tag{2.11}$$

where

$$\bar{\mathbf{X}}^T(m)\bar{e}(m) = [f_0(m), f_1(m), ..., f_{L-1}(m)]^T, \tag{2.12}$$

in which $\bar{e}(m)$ is an M element vector of errors given by

$$\bar{e}(m) = [e(mM), ..., e(mM + M - 1)]^T, \tag{2.13}$$

with $e(k) = y(k) - \hat{y}(k)$, where $y(k)$ is desired response sample. In block form $\bar{e}(m) = \bar{y}(m) - \hat{\bar{y}}(m)$. Substituting (2.8) and (2.13) into (2.12) yields the following expression for the ith element of the block mean-square error gradient estimate vector:

$$\mathbf{f}_i(m) = \mathbf{X}^T(mM - i)\bar{e}(m), \quad i = 0, 1, ..., L - 1. \tag{2.14}$$

Equation (2.14) represents a correlation between the error block (vector) $e(mM + j)$ and the shifted input data block $x(mM + j - i)$, as can be seen by writting it as a summation

$$f_i(m) = \sum_{j=0}^{M-1} x(mM + j - i)e(mM + j), \quad i = 0, 1, ..., L - 1. \tag{2.15}$$

It is apparent that the block adaptive algorithm consists of successive operations of block convolution (2.9), gradient estimation, and weight update (2.11). Note that the most efficient choice of block length is $M = L$.

2.2 Parallel LMS Adaptive Filter

We propose here an algorithm for realizing adaptive filters that achieve two objectives simultaneously. The first objective is to allow arbitrarily high sampling rates for a given speed hardware, at the expense of additional hardware. The second objective is to not modify the characteristics of the algorithm, and hence not affect the convergence and tracking capabilities.

2.2.1 Derivation of the Parallel LMS Adaptive Filter Algorithm

Consider the LMS Eq. (2.2). The solution of the Eq. (2.2) is [2]:

$$\mathbf{W}(n) = \mathbf{W}(k) + \mu \sum_{j=k}^{n-1} e(j)\mathbf{X}(j), \quad k = 0, 1, ..., \tag{2.16}$$

where $\mathbf{W}(k)$, $e(k)$, and $\mathbf{X}(k)$ are defined in (2.3) and (2.4). Substituting $n = mM + M$, and $k = mM$, $m = 0, 1...$, into (2.16), we obtain

$$\begin{aligned}\mathbf{W}(mM + M) &= \mathbf{W}(mM) + \mu \sum_{j=mM}^{mM+M-1} e(j)\mathbf{X}(j) \\ &= \mathbf{W}(mM) + \mu \sum_{j=1}^{M} e(mM + j - 1)\mathbf{X}(mM + j - 1) \\ &= \mathbf{W}(mM) + \mu \bar{\mathbf{X}}^T(m)\bar{e}(m), \quad m = 0, 1, ...,\end{aligned}$$

or

$$\bar{\mathbf{W}}(m + 1) = \bar{\mathbf{W}}(m) + \mu \bar{\mathbf{X}}^T(m)\bar{e}(m), \quad m = 0, 1, ..., \tag{2.17}$$

where

$$\bar{\mathbf{W}}(m + 1) = \mathbf{W}(mM + M), \quad \bar{\mathbf{W}}(m) = \mathbf{W}(mM),$$

$\hat{\mathbf{X}}(m)$ is an $M \times L$ matrix as in (2.8), and

$$\bar{e}(m) = [e(mM), ..., e(mM + M - 1)]^T.$$

Vector $\bar{e}(m)$ is formed using values $e(mM), ..., e(mM + M - 1)$. Now we show how to compute these values. Substituting (2.16) into (2.4), we have

$$e(n) = y(n) - \mathbf{X}^T(n)\mathbf{W}(k) - \mu \mathbf{X}^T(n) \sum_{j=k}^{n-1} e(j)\mathbf{X}(j), \quad n = 0, 1, ... \tag{2.18}$$

Defining $n = mM + i - 1$, and $k = mM$, $m = 0, 1, 2,...,M$, and using them in (2.18), we get

$$e(mM+i-1) = y(mM+i-1) - \mathbf{X}^T(mM+i-1)\mathbf{W}(mM) \tag{2.19}$$

$$- \mu\mathbf{X}^T(mM+i-1)\sum_{j=mM}^{mM+i-2} e(j)\mathbf{X}(j)$$

$$= y(mM+i-1) - \mathbf{X}^T(mM+i-1)\mathbf{W}(mM)$$

$$- \mu\mathbf{X}^T(mM+i-1)\sum_{j=1}^{i-1} e(mM+j-1)\mathbf{X}(mM+j-1)$$

or in matrix form:

$$\bar{e}(m) = \bar{y}(m) - \bar{\mathbf{X}}(m)\bar{\mathbf{W}}(m) - \bar{\mathbf{D}}(m)\bar{e}(m). \tag{2.20}$$

From (2.20), we obtain

$$\bar{e}(m) = (\mathbf{I}_M + \bar{\mathbf{D}}(m))^{-1}(\bar{y}(m) - \bar{\mathbf{X}}(m)\bar{\mathbf{W}}(m)), \quad m = 0, 1, 2, ..., \tag{2.21}$$

where $\mathbf{I}_M$ is the unity matrix,

$$\bar{y}(m) = [y(mM), ..., y(mM+M-1)]^T,$$

$$\bar{\mathbf{W}}(m) = [w_0(mM), ..., w_{L-1}(mM)]^T.$$

$$\bar{\mathbf{D}}(m) = [d_{ij}(m)], \quad i, j = 1, 2, ..., M, \tag{2.22}$$

in which

$$d_{ij}(m) = 0, \text{if } i \leq j,$$

$$d_{ij}(m) = \mu\mathbf{X}^T(mM+i-1)\mathbf{X}(mM+j-1), \quad \text{if } i > j.$$

Parallel LMS adaptive filter algorithm Eqs. (2.17) and (2.21) allows us to compute the values of weight vector $\bar{\mathbf{W}}(m) = \mathbf{W}(mM)$, $m = 0, 1, ...$ at the time instants 0, M, $2M$ and so on. In the following section, we shall derive modified parallel adaptive algorithm which allows us to compute M values of $\mathbf{W}(k)$ simultaneously.

2.2.2 *Modified Parallel LMS Adaptive Filter Algorithm 1*

Defining $n = mM+l$, and $k=mM-M+l$, $m = 0, 1,..., l = 1, 2,...,M$, and using them in (2.16) we obtain

$$\mathbf{W}(mM+l) = \mathbf{W}(mM-M+l) + \mu \sum_{j=mM-M+l}^{mM+l-1} \mathbf{X}(j)e(j)$$

$$= \mathbf{W}(mM-M+l) + \mu \sum_{j=1}^{M} \mathbf{X}(mM-M+l+j-1)$$

$$e(mM-M+l+j-1),$$

$l = 1, 2, ..., M$ or in matrix form:

$$\bar{\mathbf{W}}(m) = \bar{\mathbf{W}}(m-1) + \bar{\mathbf{B}}(m)\bar{\mathbf{E}}(m),\ m = 0, 1, ..., \tag{2.23}$$

where the $ML \times 1$ vectors $\bar{\mathbf{W}}(m)$ and $\bar{\mathbf{W}}(m-1)$ are defined by

$$\bar{\mathbf{W}}(m) = [W(mM+1), ..., W(mM+l), ..., ..., W(mM+M)]^T,$$

$$\bar{\mathbf{W}}(m-1) = [W((m-1)M+1), ..., W((m-1)M+l), ..., ..., W(mM)]^T,$$

the $ML \times M^2$ matrix $\bar{\mathbf{B}}(m)$ is defined by

$$\bar{\mathbf{B}}(m) = \texttt{blockdiag}\{\bar{B}(m,1), ..., \bar{B}(m,l), ..., ..., \bar{B}(m,M)\}, \tag{2.24}$$

in which the $L \times M$ matrix

$$\bar{\mathbf{B}}(m,l) = [X(mM-M+l), ..., X(mM+l-1)],$$

and the $M^2 \times 1$ vector $\bar{\mathbf{E}}(m)$ is defined by

$$\bar{\mathbf{E}}(m) = [e(mM-M+j), ..., e(mM-M+l+j-1), ..., e(mM+j-1)]^T \tag{2.25}$$

$$l = 1, 2, ..., M,$$

in which

$$e(mM-M+l+j-1) = [e(mM-M+l), ..., e(mM+l-1)]^T,\ \ j = 1, 2, ..., M.$$

Vector $\bar{\mathbf{E}}(m)$ in (2.23) is formed from values $e(mM), ..., e(mM+M-1)$, which we compute using Eq. (2.21). Modified parallel LMS adaptive algorithm allows us to compute M weights $W(mM+1), ..., W(mM+M)$ simultaneously. Thus, this algorithm is useful in the case we need to calculate all weights and to ensure high speed adaptive filtering.

2.2.3 *Modified Parallel LMS Adaptive Filter Algorithm 2*

Defining $n = mM+i$, $k = mM$, $m = 0, 1,..., i = 1, 2,..., M$, and using them in (2.16) we obtain

$$\mathbf{W}(mM+i) = \mathbf{W}(mM) + \mu \sum_{j=mM}^{mM+i-1} e(j)\mathbf{X}(j)$$

$$= \mathbf{W}(mM) + \mu \sum_{j=1}^{i} e(mM+j-1)\mathbf{X}(mM+j-1),\ i = 1, 2, ..., M,$$

or in matrix form:

$$\bar{\mathbf{W}}(m) = \bar{\mathbf{W}}(mM) + \mu\bar{\mathbf{X}}(m)\bar{e}(m), \tag{2.26}$$

where the $ML\times 1$ vectors $\bar{\mathbf{W}}(m)$ and $\bar{\mathbf{W}}(m-1)$ are defined by

$$\bar{\mathbf{W}}(m) = [W(mM+1), ..., W(mM+M)]^T,$$

$$\bar{\mathbf{W}}(mM) = [W(mM), ..., W(mM)]^T,$$

$$\bar{e}(m) = [e(mM), ..., e(mM+M-1)]^T,$$

and the $LM \times M$ matrix $\bar{\mathbf{X}}(m)$ is defined by

$$\bar{\mathbf{X}}(m) = [X_{ij}],\quad i, j = 1, 2, ..., M, \tag{2.27}$$

in which

$$X_{ij} = 0,\ \text{if } i < j,$$

$$X_{ij} = X(mM+j-1),\ \text{if } i \geq j.$$

Vector $\bar{e}(m) = [e(mM), ..., e(mM+M-1)]^T$ is computed using Eq. (2.20). Modified parallel LMS adaptive Algorithm 2 calculates all weights in every time instant, and enables us to compute M weights simultaneously. Thus, it is applicable in the parallel adaptive filtering schemes.

2.3 Algorithm Realization Considerations

To realize the parallel LMS adaptive algorithms, we need to calculate inverse matrix $\bar{\mathbf{T}}^{-1} = (\mathbf{I}_M + \bar{\mathbf{D}}(m))^{-1}$ see Eq. (2.21). Define

$$\bar{g}(m) = \bar{y}(m) - \bar{\mathbf{X}}(m)\bar{\mathbf{W}}(m) = [g_1(m), ..., g_M(m)]^T.$$

Then (2.21) in matrix form:

$$\begin{bmatrix} 1 & & & & \\ d_{21} & 1 & & 0 & \\ d_{31} & d_{32} & 1 & & \\ \cdot & & & \cdot & \\ d_{M,1} & \cdot & \cdot & d_{M,M-1} & 1 \end{bmatrix} \begin{bmatrix} e_1(m) \\ \cdot \\ \cdot \\ \cdot \\ e_M(m) \end{bmatrix} = \begin{bmatrix} g_1(m) \\ \cdot \\ \cdot \\ \cdot \\ g_M(m) \end{bmatrix}, \tag{2.28}$$

where $\bar{e}(m) = [e_1(m), ..., e_i(m), ..., e_M(m)]^T$, in which $e_i(m) = e(mM + i - 1)$, i = 1,2,...,M and $d_{ij}(m)$, i, j = 1,2,...,M are as in (2.22).

From (2.28) we obtain recurrent equation for calculation $e_i(m)$:

$$e_i(m) = g_i(m) - \sum_{j=1}^{i-1} d_{ij} e_j(m), \quad i = 1, 2, ..., M. \tag{2.29}$$

In some practical situations, when we need to calculate all $e_i(m)$, $i = 1, 2, ...M$ values simultaneously, the following algorithm for calculation inverse matrix is defined by

Proposition *The inverse matrix* $\bar{\mathbf{T}}^{-1}(m)$ *of the* $M \times M$ *lower matrix* $\bar{\mathbf{T}} = [d_{ij}(m)]$, $d_{ii}(m)$=*1*, i, j=*1,2,...,M is defined by*

$$\bar{\mathbf{T}}^{-1}(m) = \begin{cases} 0 & if\, i < j, \\ 1 & if\, i = j, \\ d_{ij}(m), & if\, i = j + 1, \\ \mathtt{det}(\bar{\mathbf{D}}_{ij}(m)) = [d_{kl}(m)], & \\ k = j + 1, j + 2, ..., i, & \\ l = j, j + 1, ..., (i - 1), & if\, i > j + 1. \end{cases} \tag{2.30}$$

Calculation of the matrix $\bar{\mathbf{D}}(m)$.

From (2.22) it follows for $i > j$, that

$$d_{ij}(m) = \mu \mathbf{X}^T (mM + i - 1)\mathbf{X}(mM + j - 1)$$

$$= \mu[x(mM + i - 1)x(mM + j - 1) + x(mM + i - 2)$$

$$x(mM + j - 2) + ... + x(mM + i - L)x(mM + j - L)], \tag{2.31}$$

and

$$d_{i+1,j+1}(m) = \mu \mathbf{X}^T (mM + i)\mathbf{X}(mM + j) =$$

$$\mu[x(mM + i)x(mM + j) + x(mM + i - 1)x(mM + j - 1)$$

$$+\cdots + x(mM+i-L+1)x(mM+j-L+1)]. \tag{2.32}$$

Thus, using (2.31) and (2.32), we obtain

$$d_{i+1,j+1}(m) = d_{ij}(m) + \mu[x(mM+i)x(mM+j)$$

$$-x(mM+i-L)x(mM+j-L)], \tag{2.33}$$

for j=1, 2,...,$M-2$, and $i \geq j+1$; m = 1, 2, ... Equation (2.33) provides the computations to be performed recursively along the subdiagonals of the matrix $\bar{\mathbf{D}}(m)$.

Now we derive the recursive expression of the first column $\mathbf{d}_{i,1}(mM+M)$ of matrix $\bar{\mathbf{D}}(m)$ in terms of the last row $\mathbf{d}_{M,M-i+1}(mM)$ of matrix $\bar{\mathbf{D}}(m)$ during the previous block. The first column of matrix $\bar{\mathbf{D}}(m)$ (2.22) can be expressed as:

$$\mathbf{d}_{i,1}(mM+M) = \mu\mathbf{X}^T(mM+M+i-1)\mathbf{X}(mM+M)$$

$$= \mu[x(mM+M+i-1)x(mM+M) + x(mM+M+i-2)x(mM+M-1)+\cdots$$

$$x(mM+M+i-l-1)x(mM+M-l)+...$$

$$+x(mM+M+i-L)x(mM+M-L+1)]. \tag{2.34}$$

The last row of matrix $\bar{\mathbf{D}}$ during the previous block can be expressed as:

$$\mathbf{d}_{M,M-i+1}(mM) = \mu\mathbf{X}^T(mM+M-1)\mathbf{X}(mM+M-i)$$

$$= \mu[x(mM+M-1)x(mM+M-i) + x(mM+M-2)x(mM+M-i-1)+\cdots$$

$$x(mM+M-l-1)x(mM+M-i-l)+\cdots$$

$$+x(mM+M-L)x(mM+M-L-i+1)]. \tag{2.35}$$

Then it follows from (2.34) and (2.35), that

$$\mathbf{d}_{i,1}(mM+M) = \mathbf{d}_{M,M-i+1}(mM)$$

$$= \mu[\sum_{l=0}^{i-1} x(mM+M+l)x(mM+M+l-i+1)$$

$$-\sum_{l=0}^{i-1} x(mM+M+l-L)x(mM+M+l-L-i+1)], \tag{2.36}$$

$i = 2, 3, \ldots, M; m = 0, 1, \ldots$.

So, the matrix is computed as follows:

1. Using the last row $\mathbf{d}_{M,M-i+1}(mM)$ of the matrix $\bar{\mathbf{D}}$, we compute the first column $\mathbf{d}_{i,1}(mM + M)$ of the matrix $\bar{\mathbf{D}}$ during the next block (2.36).
2. Using these first column values, we provide the computations recursively along the subdiagonals of the matrix $\bar{\mathbf{D}}$ (2.33).

2.4 Conclusions

We show that the LMS algorithm can be speeded up by the use such block formulation, which allows LMS algorithm to be exactly equivalent to the sample-by-sample LMS. We demonstrate a realization of adaptive filters for which there is no theoretical limit on sampling rate in a given speed of hadware, at the expense of additional hardware and latency. Our realization does not change the adaptive characteristics and hence does not degrade the adaptive filter tracking capability. This technique can be applied to many different algorithms in the LMS family.

References

1. Clark, G.A., Mitra, S.K., Parker, S.R.: A unified approach to time and frequency domain realization of FIR adaptive digital filters. IEEE Trans. ASSP **31**, 1073–1083 (1983)
2. Kazlauskas, K.: Pipelined-block models of linear discrete-time linear periodically time-varying systems. Informatica, LT MII VU Press Vilnius **8**(2), 215–236 (1997)
3. Widrow, B., Stearns, S.D.: Adaptive Signal Processing. Englewood Cliffs, Prentice-Hall, NJ (1985)

Chapter 3
Optimal Parametric Identification of Linear Periodically Time-Variant Systems

This chapter discusses the LPTV system parametric identification using a block approach. An block algorithm is proposed for optimal estimation of the parameters of LPTV system from the input sequence and the output sequence corrupted by additive noise with Gaussian distribution. In the proposed method the LS error criterion has been used. The algorithm provides a useful computational tool based on an appropriate theoretical foundation for parameter estimation of LTI systems from input and output data. Simulation results are presented that demonstrate the performance of the approach.

3.1 Problem Formulation

It is assumed that a LPTV system is causal, linear with an input sequence $x(k)$, $k = 0, 1, \ldots$ and an output sequence $u(k)$, $k = 0, 1, \ldots$, so that it can be described by a linear difference equation

$$u(k) = \sum_{i=0}^{N} \beta_i(k)x(k-i) - \sum_{j=1}^{M} \alpha_j(k)u(k-j),\ k = 0, 1, \ldots; \tag{3.1}$$

$N \leq M$. The measured output $y(k)$ is assumed to be contaminated with the measurement noise $n(k)$ i.e.,

$$y(k) = u(k) + n(k).$$

Equation (3.1) is L-periodic, i.e., parameters of LPTV system satisfies conditions

$$\alpha_j(k) = \alpha_j(k+L),\ \beta_i(k) = \beta_i(k+L)\ \forall k = 0, 1, \ldots;$$

R. Pupeikis and K. Kazlauskas, *Advanced Digital Signal Processing Methods for Filtering, Identification, and Nonlinear Systems Control*, Studies in Systems, Decision and Control 638, https://doi.org/10.1007/978-3-032-06740-1_3

L is periodicity, $L = 2, 3, \ldots$

A noise $n(k)$ is a nonmeasurable, normally distributed statistically independent(discrete white noise) with $\mathcal{E}\{n(k)\} = 0$, $\mathcal{E}\{n(k)n(k+\tau)\} = \sigma_n^2\delta(\tau)$, where $\mathcal{E}\{n(k)\}$ is a mean value, σ_n^2 is the variance and $\delta(\tau)$ is the Kronecker delta function. The basis of LPTV parameter estimation is the set of data $\{x(0), x(1), \ldots, x(T), y(0), y(1), \ldots, y(T)\}$ of observations of the input sequence $\{x(k)\}$ and the noisy output sequence $\{y(k)\}$, $k = 0, 1, \ldots, T$.

The purpose of the chapter is to investigate the block parameter estimation approach in the case of additive noise $n(k)$, $k = 0, 1, \ldots$, acting on the output of the LPTV system (3.1) to be estimated.

3.2 Block Model

The objective of parameter estimation is to estimate the LPTV system parameters in (3.1) based on measured sequences $x(k)$ and $y(k)$. It is assumed that the M and N are known a priori. If this is not the case they can be determined by order search methods [2]. The noise $n(k)$ is assumed to be stationary.

Consider that the input sequence $x(k)$ and the output sequence $y(k)$ of the LPTV system are measured for $k = 0, 1, \ldots, T$, The model of the LPTV system can be described by a linear difference equation

$$y(k) + \sum_{j=1}^{M} a_j(k)y(k-j) - \sum_{i=0}^{N} b_i(k)x(k-i) = e(k),$$

$k = 0, 1, \ldots;\ N \leq M$, where

$$a_j(k) = a_j(k+L),\ \ b_i(k) = b_i(k+L)\ \forall k = 0, 1, \ldots;$$

are parameters of the model. The equation error (residual) $e(k)$ arises from the noise contaminated output $y(k)$ and from erroneous parameter estimates.

Replacing k by $l + (m-1)L$ and blocking the input sequence $x(k)$ and the output sequence $y(k)$, we obtain the equation of the block model

$$y(l + (m-1)L) + \sum_{j=1}^{M} a_j(l)y(l + (m-1)L - j)$$

$$- \sum_{i=0}^{N} b_i(l)x(l + (m-1)L - i) = e(l + (m-1)L),$$

or

$$y(l+(m-1)L) = \sum_{i=0}^{N} b_i(l)x(l+(m-1)L-i)$$

$$-\sum_{j=1}^{M} a_j(l)y(l+(m-1)L-j) + e(l+(m-1)L), \tag{3.2}$$

where $l = M+1,..., M+L;\ m = 1, 2,..., p;\ p = \texttt{fix}((T-M-1)/L)$ rounds the $(T-M-1)/L$ to the nearest integer towards zero.

The system of linear equations (3.2) can be expressed in a matrix form as follows:

$$\bar{y}(l) = \mathbf{A}(l)\theta(l) + \bar{e}(l)\ \ l = M+1, ..., M+L, \tag{3.3}$$

where

$$\bar{y}(l) = [y(l), y(l+L), ..., y(l+(m-1)L), ..., y(l+(p-1)L)]^T \tag{3.4}$$

is a p-element vector,

$$\theta(l) = [b_0(l), ..., b_i(l), ..., b_N(l), a_1(l), ..., a_j(l), ..., a_M(l)]^T \tag{3.5}$$

is a $(N+M+1)$-element vector,

$$\bar{e}(l) = [e(l), e(l+L), ..., e(l+(m-1)L), ..., e(l+(p-1)L)]^T \tag{3.6}$$

is a p-element vector,

$$\mathbf{A}(l) = \begin{bmatrix} x(l), ..., x(l-i), ..., x(l-N), -y(l-1), ..., -y(l-j), ..., -y(l-M) \\ \cdots \\ x(l+(m-1)L), ..., x(l+(m-1)L-i), ..., x(l+(m-1)L-N), \\ -y(l+(m-1)L-1), ..., -y(l+(m-1)L-j), ..., -y(l+(m-1)L-M), \\ \cdots \\ x(l+(p-1)L), ..., x(l+(p-1)L-i), ..., x(l+(p-1)L-N), \\ -y(l+(p-1)L-1), ..., -y(l+(p-1)L-j), ..., -y(l+(p-1)L-M), \end{bmatrix} \tag{3.7}$$

is a $(p \times (N+M+1))$ data matrix.

3.3 Nonrecursive Estimation Procedure

Define $x_l(m) = x(l+(m-1)L)$ and $y_l(m) = y(l+(m-1)L)$, then errors of the block model (3.2)

$$e_l(m) = y_l(m) + \sum_{j=1}^{M} a_j(l) y_l(m-j) - \sum_{i=0}^{N} b_i(l) x_l(m-i),$$

$$m = 1, 2, ...; l = M+1, ..., M+L, \tag{3.8}$$

where $e_l(m) = e(l + (m-1)L)$.

Minimization of the loss functions

$$V_l = \min_{\theta(l)} \{\bar{e}^T(l)\bar{e}(l)\}$$

$$= \sum_{m=1}^{p} e^2(l + (m-1)L), \quad l = M+1, ..., M+L, \tag{3.9}$$

and therefore

$$\frac{\partial V_l}{\partial \theta(l)} |_{\theta(l)=\hat{\theta}(l)} = 0, \quad l = M+1, ..., M+L$$

results in the optimal parameter estimation. Assuming that **A** is full rank, the LS solution of the overdetermined system of linear equations (3.2) is

$$\hat{\theta}(l) = (\mathbf{A}^T(l)\mathbf{A}(l))^{-1}\mathbf{A}^T(l)\bar{\mathbf{y}}(l), \quad l = M+1, ..., M+L, \tag{3.10}$$

where

$$\hat{\theta}(l) = [\hat{b}_0(l), ..., \hat{b}_i(l), ..., \hat{b}_N(l), \hat{a}_1(l), ..., \hat{a}_j(l), ..., \hat{a}_M(l)]^T.$$

The parameter estimation problem has been transformed into an one-dimensional minimization problem (3.9) for L sets parameters of the LPTV system. This is non-recursive parameter estimation as parameter estimates are obtained only after measuring and storing of all signal values. Apart from using the standard LS approach the overdetermined system of equations (3.3) can be solved using the total LS method [1]. This method assumes that there are estimation errors in the elements of both **A** and $\bar{\mathbf{y}}$. **A** and $\bar{\mathbf{y}}$ are then modified so that the rank of the extended matrix $[\mathbf{A}\ \bar{\mathbf{y}}]$ equals to rank of **A**.

Using the backward-shift operator z^{-1} such that $z^{-1}y(k) = y(k-1)$, the block model (3.2) can be described in the form

$$e_l(m) = A_l(z^{-1})y_l(m) - B_l(z^{-1})x_l(m),$$

$$l = M+1, ..., M+L, \ m = 1, 2, ..., \tag{3.11}$$

where

$$A_l(z^{-1}) = 1 + a_1(l)z^{-1} + \cdots + a_M(l)z^{-M},$$

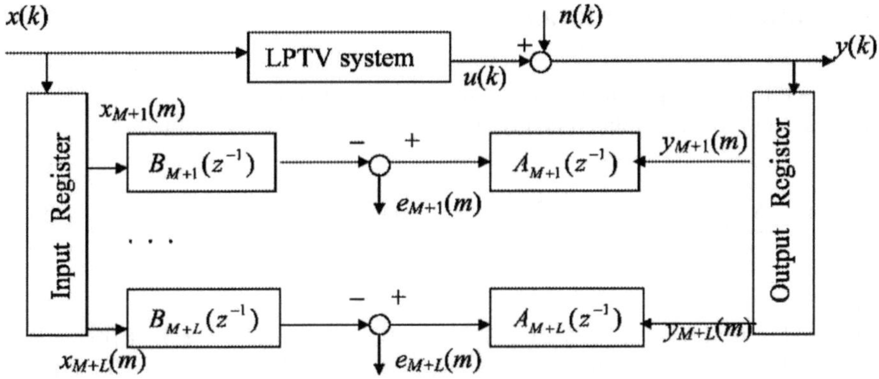

Fig. 3.1 The block estimation of the LPTV system parameters

$$B_l(z^{-1}) = b_0(l) + b_1(l)z^{-1} + \cdots + b_N(l)z^{-N}$$

Operators $A_l(z^{-1})$ and $B_l(z^{-1})$ depend on parameters $a_i(l)$ and $b_j(l)$. The errors $e_l(m)$ of the block model are linear functions of the parameters.

The block estimation of the LPTV system parameters is shown in Fig. 3.1. The input and the output registers reduce (decimate) the sampling rate of the input sequence $x(k)$ and of the output sequence $y(k)$. The sampling rate reduction is achieved by forming the subsequences $x_{M+i}(m)$ and $y_{M+i}(m)$ $i = 1, 2,..., L$ and by saving every Lth sample of the sequences $x(k)$ and $y(k)$. The decimation factor is equal to the periodicity of the LPTV system, i.e., to the block length L of sequences $x(k)$ and $y(k)$. The subsequences $x_{M+i}(m)$ and $y_{M+i}(m)$, $i = 1, 2,..., L$ at the outputs of the registers are used for the parameter estimation. The parameter estimation we begin at the time moment $l = M + 1$, where M is the order of the LPTV system. The estimation procedure can be accomplished in a parallel manner for $l = M + 1,..., M + L$ (see Fig. 3.1).

In case the LPTV system periodicity L is unknown, we can repeat the parameter estimation process with different periodicity values $L^* = 2, 3$, and so on. If $L^* \neq L$ (L is true system periodicity), then we obtain large values of loss functions (3.9) and bad parameter estimates, because the elements of the data matrix $\mathbf{A}(l)$ are not such as they must be in (3.7).

In case $L^* = L$ we obtain small values of loss functions (3.9) and optimal parameter estimates because the elements of the data matrix $\mathbf{A}(l)$ are just like they must be in (3.7). Then, by repeating the estimation process with different values of the periodicity L^*, calculating loss functions (3.9), and comparing these values, we can estimate the true value of periodicity L.

3.4 Numerical Simulation

In this section we apply the described method to an example of LPTV system. The LPTV system to be simulated is described by a linear difference equation

$$y(k) = \sum_{i=0}^{2} \beta_i(k)x(k-i) - \sum_{j=1}^{2} \alpha_j(k)y(k-j) + \lambda n(k),\ k = 0, 1, ..., \tag{3.12}$$

where
$\beta_0(0) = 1,\ \beta_0(1) = 1.2,\ \beta_0(2) = 1.5,\ \beta_1(0) = 2,\ \beta_1(1) = 2.2,\ \beta_1(2) = 2.5,$ $\beta_2(0) = 3,\ \beta_2(1) = 3.2,\ \beta_2(2) = 3.5, \alpha_1(0) = -1.1197,\ \alpha_1(1) = -1.5,\ \alpha_1(2) =$ $-1.036,\ \alpha_2(0) = 0.3012,\ \alpha_2(1) = 0.7,\ \alpha_2(2) = 0.2636,$

The periodicity L of the LPTV system (3.1) is equal to 3. The noise sequence $n(k)$ is generated as a zero-mean, unit-variance white noise sequence with a Gaussian disribution. λ is a constant that determines the intensity of additive noise $n(k)$. The sequences, including that of the noise $n(k)$, $k = 0, 1, \ldots, 500$ where generated using the MATLAB functions.

For calculation of the LPTV system output we used MATLAB m-file PERIOD. The matrix (3.7) was formed by using m-file MATRICA. The performance of the estimated model is summarized in Table 3.1, in which SNR=20$\log_{10}\frac{\sigma_u}{\sigma_n}$, where σ_u is standard deviation of the LPTV system output without noise, σ_n-standard deviation of the noise $n(k)$. In Table 3.1 α and β are true parameters of the LPTV system and $\hat{a}$ and $\hat{b}$ are averaged estimated parameters of the model.

For calculation of averaged estimated values $\hat{a}$ and $\hat{b}$ we repeated estimation procedure (3.10) 500 times, obtained 500 sets of estimated values, then calculated mean and standard deviation of the estimated values for SNR = 43, 28, 22, 16 and 12 dB. In Fig. 3.2 we show how the estimation quality of the LPTV system parameters depends on the number of processed observations for SNR = 12 dB. We defined a measure of the parameter estimation quality as:

$$B0 = \frac{|\ \beta_0(0) - \hat{b}_0(0)\ |}{|\ \beta_0(0)\ |} \cdot 100\%,\ A0 = \frac{|\ \alpha_1(0) - \hat{a}_1(0)\ |}{|\ \alpha_1(0)\ |} \cdot 100\%, \tag{3.13}$$

and so on, where $\hat{b}_0(0)$, $\hat{a}_1(0)$ denote averaged estimated parameters and $\beta_0(0)$, $\alpha_1(0)$ denote true parameters of the LPTV system at time moment 0. The impulse responses of LPTV system (full lines) and impulse responses of the estimated system (dotted lines) for time moments 0, 1 and 2 in case SNR = 12 dB are shown in Fig. 3.3. In Fig. 3.4 we show MFRs of LPTV system (full lines) and MFRs of the estimated system (dotted lines) for 0,1, and 2 time moments in case SNR = 12 dB.

Table 3.1 LPTV system parameter estimation using block model. True parameters and averaged estimated parameters of the LPTV system (3.12) via SNR(dB)

True parameter values	$\beta_0(0)$	$\beta_1(0)$	$\beta_2(0)$	$\alpha_1(0)$	$\alpha_2(0)$	SNR
	1	2	3	−1.1197	0.3012	(dB)
Estimated values	$b_0(0)$	$b_1(0)$	$b_2(0)$	$a_1(0)$	$a_2(0)$	
Mean	1.0026	2.0026	3.0011	−1.1188	0.3004	43
Std.	0.0341	0.0379	0.0322	0.0071	0.0072	
Mean	1.0082	2.0281	3.0510	−1.0997	0.2827	28
Std.	0.1267	0.1929	0.1658	0.0453	0.0370	
Mean	1.0109	2.1205	3.1723	−1.0473	0.2346	22
Std.	0.2510	0.2874	0.2963	0.0557	0.0544	
Mean	0.9939	2.3361	3.5897	−0.9072	0.1060	16
Std.	0.4370	0.5232	0.7051	0.1295	0.1138	
Mean	0.9995	2.5328	3.9647	−40.7579	−0.0151	12
Std.	0.7933	0.9071	0.8420	0.1444	0.1324	
True parameter values	$\beta_0(0)$	$\beta_1(0)$	$\beta_2(0)$	$\alpha_1(0)$	$\alpha_2(0)$	SNR
	1.2	2.2	3.2	−1.5	0.7	(dB)
Estimated values	$b_0(0)$	$b_1(0)$	$b_2(0)$	$a_1(0)$	$a_2(0)$	
mean	1.1983	2.1976	3.2006	−1.4990	0.6997	43
std	0.0754	0.0824	0.0455	0.0164	0.0168	
mean	1.2032	2.2618	3.2880	−1.4625	0.6654	28
std	0.1890	0.1957	0.2112	0.0432	0.0429	
mean	1.1739	2.3298	3.4876	−1.3712	0.5829	22
std	0.3531	0.3460	0.3442	0.0814	0.0778	
mean	1.1687	2.6061	4.0788	−1.1184	0.3461	16
std	0.7193	0.5577	0.9445	0.1850	0.1556	
mean	1.2662	2.8081	4.6689	−0.8914	0.1482	12
std	0.8800	0.9134	1.2331	0.1792	0.1676	
True parameter values	$\beta_0(0)$	$\beta_1(0)$	$\beta_2(0)$	$\alpha_1(0)$	$\alpha_2(0)$	SNR
	1.5	2.5	3.5	-1.036	0.264	(dB)
Estimated values	$b_0(0)$	$b_1(0)$	$b_2(0)$	$a_1(0)$	$a_2(0)$	
mean	1.5010	2.5005	3.5008	−1.0358	0.2635	43
std	0.0232	0.0248	0.0335	0.0052	0.0054	
mean	1.5100	2.5125	3.5353	−1.0239	0.2507	28
std	0.1327	0.1607	0.1407	0.0221	0.0228	
mean	1.4853	2.5572	3.6224	−0.9950	0.2248	22
std	0.2524	0.2484	0.3270	0.0433	0.0447	
mean	1.4563	2.8607	4.2015	−0.7816	0.0259	6
std	0.5572	0.5645	0.5186	0.0815	0.0803	
mean	1.4563	2.8607	4.2015	−0.7816	−0.0259	12
std	0.6437	0.7949	0.8443	0.1068	0.1050	

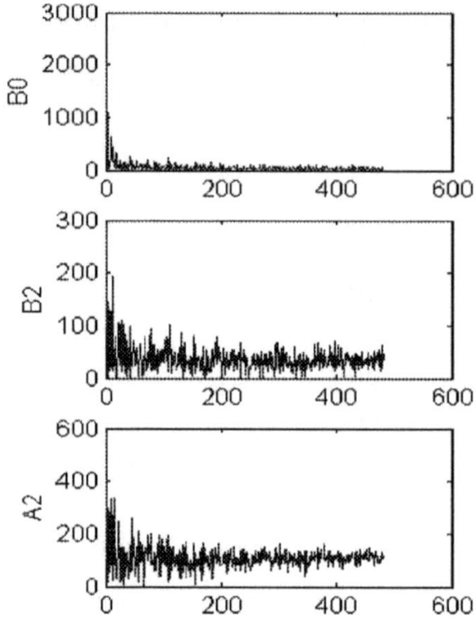

Fig. 3.2 The quality of the LPTV system parameter estimation vs the number of processed observations. SNR = 12 dB

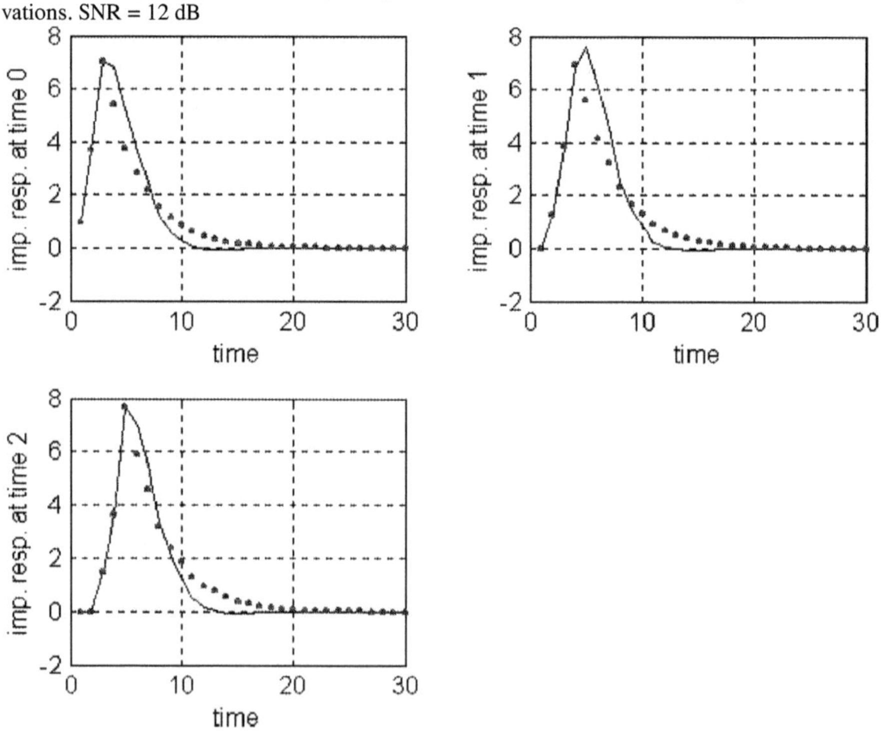

Fig. 3.3 Impulse responses of the LPTV system (3.12) (full lines) and the impulse responses of the estimated LPTV system (dotted lines). SNR = 12 dB

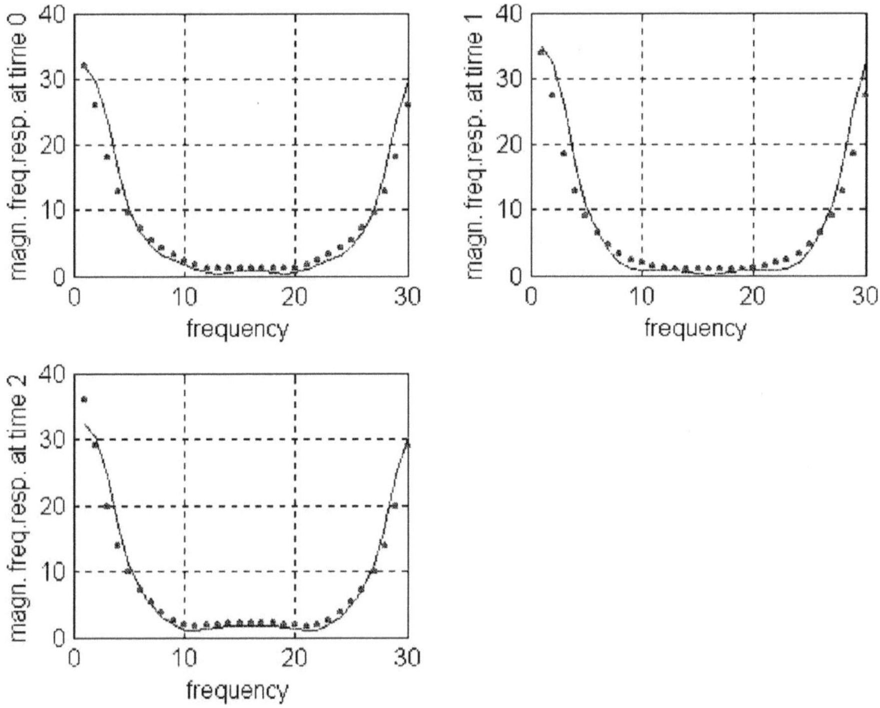

Fig. 3.4 MFRs of the LPTV system (3.12) (full lines) and MFRs of the estimated LPTV system (dotted lines). SNR = 12 dB

3.5 Conclusions

We have discussed a block parameter estimation method for linear periodically time-varying systems. Its performance strongly depends on the noise level and does not provide satisfactory parameter estimation results when the noise level is high. The estimated values are seen to be close to the true parameter values at high SNR. This property can be confirmed by the analysis of the simulation results of Table 3.1. We show that the parallel parameter estimation structure can be realized. If periodicity $L=1$, the block method reduces to the conventional LTI system parameter estimation method. The results of numerical simulation of the LPTV system by computer show efficiency of the proposed approach.

References

1. Golub, G., Van Loan, C.: Matrix Computations. MD, Johns Hopkins University Press, Baltimore (1983)
2. Isermann, R.: Prozessidentifikation. Springer, Berlin (1974)

Chapter 4
Iterative Parametric Identification Algorithm of Autoregression

The chapter considers the estimation algorithm of AR parameters in the presence of white observation noise. The estimation algorithm is iterative because the estimation of noise variance and parameters is sequentially repeated until some convergence criterion is satisfied. The algorithm is built upon the bias correction principle [3, 5, 6]. The idea is to use the HO(Y-W) equations (i.e., to use more autocorrelation samples), sequentially estimate the noise variance, and to use these variance estimates for the bias correction.

4.1 Preliminaries

Let the signal $x(t)$ be generated by the p-th order equation

$$x(t) = \sum_{i=1}^{p} a_i x(t-i) + v(t), \tag{4.1}$$

where $v(t)$ is white driving noise of zero mean and variance σ_v^2(i.e., $\mathcal{E}v(t) = 0$, $\mathcal{E}v(t)v(s) = \sigma_v^2\delta_{t,s}$), and $\{a_i\}$ are real coefficients; $\mathcal{E}$ is the mathematical expectation.

Let $y(k)$ denote the noise corrupted measurement of $x(t)$,

$$y(t) = x(t) + w(t), \tag{4.2}$$

where the noise $w(t)$ is a stationary process of zero mean and variance σ_w^2 uncorrelated with $v(t)$, N is the number of data samples.

For the noiseless case, $\{a_i, i = 1, 2, ..., p\}$ can be found using Y-W equations

R. Pupeikis and K. Kazlauskas, *Advanced Digital Signal Processing Methods for Filtering, Identification, and Nonlinear Systems Control*, Studies in Systems, Decision and Control 638, https://doi.org/10.1007/978-3-032-06740-1_4

$$r_x(k) = \sum_{l=1}^{p} a_l r_x(k-l), \; k \geq l, \tag{4.3}$$

where $r_x(k) = \mathcal{E}\{x(t)x(t-k)\}$ is the autocovariance function of $x(t)$. Clearly, any p equations are sufficient to determine the AR parameters. Generally $k = 1, 2, ..., p$ is chosen which results in a set of symmetric Toeplitz equations

$$\begin{pmatrix} r_x(0) & r_x(1) & ... & r_x(p-1) \\ r_x(1) & r_x(0) & ... & r_x(p-2) \\ ... & ... & ... & ... \\ r_x(p-1) & r_x(p-2) & ... & r_x(0) \end{pmatrix} \begin{pmatrix} a_1 \\ a_2 \\ ... \\ a_p \end{pmatrix} = \begin{pmatrix} r_x(1) \\ r_x(2) \\ ... \\ r_x(p) \end{pmatrix} \tag{4.4}$$

or in matrix form

$$\mathbf{R}_x \mathbf{a} = \mathbf{r}_x. \tag{4.5}$$

When the noise is present, we cannot estimate $r_x(k)$ since only $y(t)$ is available. In such a case

$$\begin{pmatrix} r_y(0) & r_y(1) & ... & r_y(p-1) \\ r_y(1) & r_y(0) & ... & r_y(p-2) \\ ... & ... & ... & ... \\ r_y(p-1) & r_y(p-2) & ... & r_y(0) \end{pmatrix} \begin{pmatrix} a'_1 \\ a'_2 \\ ... \\ a'_p \end{pmatrix} = \begin{pmatrix} r_y(1) \\ r_y(2) \\ ... \\ r_y(p) \end{pmatrix} \tag{4.6}$$

or in matrix form

$$\mathbf{R}_y \mathbf{a}_{ls} = \mathbf{r}_y. \tag{4.7}$$

But

$$r_y(k) = \begin{cases} r_x(0) + \sigma_w^2 & \text{if } k = 0 \\ r_x(k), & \text{if } k \neq 0 \end{cases} \tag{4.8}$$

Then, from (4.8) it follows that $r_x = r_y$ and

$$\mathbf{R}_x = \mathbf{R}_y - \sigma_w^2 \mathbf{I}_p, \tag{4.9}$$

where $\mathbf{I}_p$ is the $(p \times p)$ identity matrix.

Substituting (4.9) into (4.5), we have

$$\mathbf{a} = \mathbf{a}_{ls} + \sigma_w^2 \mathbf{R}_y^{-1} \mathbf{a}, \tag{4.10}$$

where $\mathbf{a}_{ls} = \mathbf{R}_y^{-1} \mathbf{r}_y$ is the LS estimate [2].

On the other hand, since (4.8) is valid, we may estimate $r_x(k)$ for all $k \neq 0$ from $y(t)$, since it is assumed σ_w^2 to be unknown. Then, by choosing $k = p + 1, ..., 2p$ in (4.3) and using (4.8), the resulting equations will not involve $r_x(0)$, i.e.,

$$\begin{pmatrix} r_y(p) & r_y(p-1) & \ldots & r_y(1) \\ r_y(p+1) & r_y(p) & \ldots & r_y(2) \\ \ldots & \ldots & \ldots & \ldots \\ r_y(2p-1) & r_y(2p-2) & \ldots & r_y(p) \end{pmatrix} \begin{pmatrix} a_1 \\ a_2 \\ \ldots \\ a_p \end{pmatrix} = \begin{pmatrix} r_y(p+1) \\ r_y(p+2) \\ \ldots \\ r_y(2p) \end{pmatrix} \tag{4.11}$$

or in matrix form

$$\mathbf{R}_h \mathbf{a}_h = \mathbf{r}_h. \tag{4.12}$$

These equations are called the high-order HO(Y-W) equation [4]. The HO(Y-W) estimate

$$\mathbf{a}_h = \mathbf{R}_h^{-1} \mathbf{r}_h. \tag{4.13}$$

4.2 The Iterative Estimation Algorithm

As shown in (4.10) (see also [3]), an asymptotic bias of the LS estimate

$$\mathbf{a} - \mathbf{a}_{ls} = \sigma_w^2 \mathbf{R}_y^{-1} \mathbf{a} \tag{4.14}$$

depends on the noise variance σ_w^2 and is direct proportional to the parameter vector $\mathbf{a}$. From (4.10), we obtain

$$\mathbf{a} = (\mathbf{I} - \sigma_w^2 \mathbf{R}_y^{-1})^{-1} \mathbf{a}_{ls}. \tag{4.15}$$

Keeping in mind that [1]

$$(\mathbf{I} - \mathbf{A})^{-1} = \sum_{i=0}^{\infty} \mathbf{A}^i, \tag{4.16}$$

where $\mathbf{A}$ is a $(p \times p)$ matrix, (4.15) may be written as

$$\mathbf{a} = \mathbf{a}_{ls} + \sigma_w^2 \mathbf{R}_y^{-1} \mathbf{a}_{ls} + (\sigma_w^2 \mathbf{R}_y^{-1})^2 \mathbf{a}_{ls} + \cdots \tag{4.17}$$

Define

$$\mathbf{a}^{(0)} = \mathbf{a}_{ls}, \tag{4.18}$$

$$\mathbf{a}^{(1)} = \mathbf{a}_{ls} + \sigma_w^2 \mathbf{R}_y^{-1} \mathbf{a}_{ls}, \tag{4.19}$$

$$\mathbf{a}^{(2)} = \mathbf{a}_{ls} + \sigma_w^2 \mathbf{R}_y^{-1} \mathbf{a}_{ls} + (\sigma_w^2 \mathbf{R}_y^{-1})^2 \mathbf{a}_{ls}, \tag{4.20}$$

and so on.

Next, we obtain from (4.20) the iterative parameter estimation algorithm

$$\mathbf{a}^{(i)} = \mathbf{a}_{ls} + \sigma_w^2 \mathbf{R}_y^{-1} \mathbf{a}^{i-1}, \tag{4.21}$$

which depends on the noise variance σ_w^2.

The HO(Y-W) estimate $\mathbf{a}_h$ (4.13) and the estimate $\mathbf{a} = \mathbf{R}_x^{-1}\mathbf{r}_x$ do not depend on the noise variance. On the other hand, the LS estimate a_{ls} (4.10) depends on the noise variance. So, the difference of the HO(Y-W) estimate and the LS estimate is proportional to the noise variance. Substituting $\mathbf{a}_h$ into the left side of (4.10), we get

$$\mathbf{a}_h - \mathbf{a}_{ls} = \sigma_w^2 \mathbf{R}_y^{-1}\mathbf{a}, \tag{4.22}$$

From (4.22), we obtain the estimate of the noise variance in the LS sense

$$\sigma_w^2 = \frac{(\mathbf{R}_y^{-1}\mathbf{a})^T(\mathbf{a}_h - \mathbf{a}_{ls})}{||\ \mathbf{R}_y^{-1}\mathbf{a}\ ||}, \tag{4.23}$$

where $||\cdot||$ is the l_2 norm of the vectors.

We can show that our algorithm is more general as compared with [6]. Multiplying both sides of (4.22) by $\mathbf{R}_h$, we obtain

$$\mathbf{r}_h - \mathbf{R}_h\mathbf{a}_{ls} = \sigma_w^2 \mathbf{R}_h \mathbf{R}_y^{-1}\mathbf{a}. \tag{4.24}$$

From (4.24), it follows

$$\sigma_w^2 = \frac{\mathbf{r}_h - \mathbf{R}_h\mathbf{a}_{ls}}{\mathbf{R}_h\mathbf{R}_y^{-1}\mathbf{a}}. \tag{4.25}$$

In case we use only one additional point $r_y(p+1)$ of the autocovariance function, then in (4.12)

$$\mathbf{R}_h = [r_y(p) r_y(p-1) ... r_y(1)], \quad r_h = r_y(p+1),$$

and (4.25) is equal to (4.24) in [6].

The Iterative Algorithm:

Using noisy samples $y(1)$, ..., $y(N)$ compute autocovariance estimates $\hat{r}_y(k)$, k=0, 1,..., $2p$, form the autocovariance matrix estimate $\hat{\mathbf{R}}_y$ (4.7), vector estimate $\hat{\mathbf{r}}_y$ (4.7), autocovariance matrix estimate $\hat{\mathbf{R}}_h$ (4.12), and the vector estimate $\hat{\mathbf{r}}_h$ (4.12).

1. Calculate the LS estimate (4.7)

$$\hat{\mathbf{a}}^{(0)} = \hat{\mathbf{a}}_{ls} = \hat{\mathbf{R}}_y^{-1}\hat{\mathbf{r}}_y \tag{4.26}$$

and the H-OY-W estimate (4.13)

$$\hat{\mathbf{a}}_h = \hat{\mathbf{R}}_h^{-1}\hat{\mathbf{r}}_h. \tag{4.27}$$

2. Estimate the noise variance (4.23)

$$\hat{\sigma}_w^{2(i)} = \frac{(\hat{\mathbf{R}}_y^{-1}\hat{\mathbf{a}}^{(i-1)})^T(\hat{\mathbf{a}}_h - \hat{\mathbf{a}}_{ls})}{\| \hat{\mathbf{R}}_y^{-1}\hat{\mathbf{a}}^{(i-1)} \|}. \tag{4.28}$$

3. Use the parameter estimation algorithm (4.21) to obtain

$$\hat{\mathbf{a}}^{(i)} = \hat{\mathbf{a}}_{ls} + \hat{\sigma}_w^{2(i)}\hat{\mathbf{R}}_y^{-1}\hat{\mathbf{a}}^{(i-1)}. \tag{4.29}$$

4. Terminate the calculations, if

$$\frac{\| \hat{\mathbf{a}}^{(i)} - \hat{\mathbf{a}}^{(i-1)} \|}{\| \hat{\mathbf{a}}^{(i)} \|} < \varepsilon, \tag{4.30}$$

where ε is a positive number; otherwise set the iteration step $i = i{+}1$ and go to 3.

4.3 Simulation Results

Simulation results are given to illustrate the proposed iterative estimation algorithm. We analyze the fourth order noisy AR process (4.1), (4.2), where [6]: $\mathbf{a}^T$=[1.352, –1.338, 0.662, –0.24], σ_v^2=1.0.

The measure of the signal to noise ratio is

$$SNR = 10\mathtt{log}\frac{\sigma_x^2}{\sigma_w^2}\ (\mathtt{dB}). \tag{4.31}$$

The stop criterion for iterative algorithm is chosen

$$\frac{\| \hat{\mathbf{a}}^{(i)} - \hat{\mathbf{a}}^{(i-1)} \|}{\| \hat{\mathbf{a}}^{(i)} \|} < \varepsilon, \tag{4.32}$$

where $\hat{\mathbf{a}}^{(i)}$ is the parameter estimate at the ith iteration step; $\varepsilon = 10^{-4}$.

The parameter estimation error is

$$\frac{\| \mathtt{mean}(\hat{\mathbf{a}}) - \mathbf{a} \|}{\| \mathbf{a} \|} \cdot 100\%, \tag{4.33}$$

where $\mathbf{a}$ is the true parameter vector, $\mathtt{mean}(\hat{\mathbf{a}})$ is the mean of the parameter vector estimates in 150 Monte Carlo tests.

We consider the performance of the iterative algorithm for different SNR when noisy observations N=6000 of $y(t)$ are available. Tables 4.1, 4.2 and 4.3 show that the LS method is very sensitive to the measurement noise, while the improved LS

Table 4.1 MonteCarlo simulation results: true parameters and averaged estimated parameters of the noisy AR process via $SNR \approx 15$dB, $\sigma_w^2 = 0.122$, $N = 6000$

True value	LS	ILSNP	ILSD	Iterative
$a_1 = 1.352$	1.077± 0.016	1.351± 0.042	1.351± 0.042	1.351± 0.040
$a_2 = -1.338$	–0.872± 0.024	–1.330± 0.071	–1.330± 0.070	–1.3340±0.065
$a_3 = 0.662$	0.230± 0.024	0.659± 0.063	0.658± 0.063	0.659± 0.059
$a_4 = -0.240$	–0.040± 0.016	-0.239± 0.030	–0.237± 0.030	–0.240± 0.002
$\hat{\sigma}_w^2$	–	0.121± 0.017	0.121± 0.017	0.127± 0.013
Error	12.588%	0.001%	0.001%	≈ 0.000%

Table 4.2 Marks and notation are the same as in Table 4.1, except of distinct values $SNR \approx 11$dB, $\sigma_w^2 = 0.3$, $N = 6000$

True value	LS	ILSNP	ILSD	Iterative
$a_1 = 1.352$	0.887± 0.016	1.355± 0.081	1.354± 0.082	1.354± 0.056
$a_2 = -1.338$	–0.594± 0.022	–1.348± 0.129	–1.347± 0.130	–1.340± 0.091
$a_3 = 0.662$	0.010± 0.021	0.673± 0.116	0.673± 0.118	0.664± 0.080
$a_4 = -0.240$	0.032± 0.015	–0.245± 0.051	–0.245± 0.056	–0.238± 0.039
$\hat{\sigma}_w^2$	–	0.293± 0.034	0.293± 0.034	0.297± 0.030
Error	30.832%	0.006%	0.006%	≈0.000%

Table 4.3 Marks and notation are the same as in Table 4.1, except of distinct values $SNR \approx 7.0$dB, $\sigma_w^2 = 0.7573$, $N = 6000$

True value	LS	ILSNP	ILSD	Iterative
$a_1 = 1.352$	0.668± 0.014	1.252± 0.256	1.271± 0.263	1.301± 0.141
$a_2 = -1.338$	–0.364± 0.016	–1.217± 0.361	–1.221± 0.369	–1.241± 0.181
$a_3 = 0.662$	–0.137± 0.016	0.527± 0.291	0.531± 0.290	0.601± 0.157
$a_4 = -0.240$	0.045± 0.012	–0.203± 0.098	–0.204± 0.191	–0.220± 0.067
$\hat{\sigma}_w^2$	–	0.521± 0.361	0.513± 0.350	0.710± 0.220
Error	51.872%	1.071%	0.939%	0.388%

method with no prefiltering (ILSNP) [5], the improved LS method with direct structure implementation (ILSD) [6] and iterative estimation algorithms yield consistent estimates of the noise variance and AR parameters. On the other hand, we see that the iterative estimation algorithm is better than ILSNP and ILSD methods in terms of standard deviations of the estimated parameters, i.e., standard deviations of the estimates of the iterative algorithm are less as compared with the ILSNP and ILSD methods. It is due to the fact that the iterative algorithm requires more additional autocovariance points $r_y(p), r_y(p+1), ..., r_y(2p)$, i.e., more information about the noisy signal $y(t)$ is used for the evaluation of noise variance.

4.4 Conclusions

This chapter introduces a new iterative estimation algorithm for system parametric identification. The noise compensation approach presented in the paper offers a better method than the ARMA method for reducing the white observation noise effects on the parameter estimates. The bias error of the LS method introduced by noise can be reduce. An advantage of the iterative algorithm is its ability to get more accurate estimates of the AR parameters as compared with the ILSNP and ILSD methods. The use of the HO(Y-W) equations (4.12) is equivalent to noise compensation in the autocovariance matrix (4.7). We approximate a set of the first $2p+1$ autocovariance estimates to a noisy AR process of order p. Our approach is more general as compared with the ILSNP method [6], because we use more additional samples of the autocovariance function for the noise variance estimation.

References

1. Bellman, R.: Introduction to Matrix Analysis. McGraw-Hill Book Company, NI (1960)
2. Davis, M.H.A., Vinter, R.B.: Stochastic Modeling and Control. Chapman and Hall, London (1985)
3. Stoica, P., Nehorai, A., Kay, S.M.: Statistical analysis of the least squares autoregressive estimator in the presence of noise. IEEE Trans. ASSP **35**(9), 1273–1281 (1987)
4. Vergara-Dominiquez, L.: New insights into the high-order Yule-Walker equations. IEEE Trans. ASSP **38**(9), 1649–1651 (1990)
5. Zheng, W.X.: A least-squares based method for autoregressive signals in the presence of noise. IEEE Trans. Circuits Syst. II **46**(1), 81–85 (1999)
6. Zheng, W.X.: Autoregressive parameter estimation from noisy data. IEEE Trans. Circuits Syst. **47**(1), 71–75 (2000)

Chapter 5
Layered Polynomial Filter Structures

It is shown that nonlinear Volterra, polynomial AR, and bilinear filters have the same layered implementation procedure. Using the layered structure, the order of nonlinearity can be increased by adding more layers to the structure. This chapter develops a unified, computationally efficient and modular structures for polynomial nonlinear filter implementation using the layered approach. The layered approach is related to Priestley's state dependent linear models [5].

5.1 Overview of Polynomial Filter Implementation

Volterra filters can be realized as linear filters by changing the nonlinear problem into a linear one. For example, the second order Volterra filter

$$y(n) = \sum_{i_1=0}^{N} a(i_1)u(n-i_1) + \sum_{i_1=0}^{N}\sum_{i_2=0}^{N} a(i_1,i_2)u(n-i_1)u(n-i_2) \tag{5.1}$$

can be realized as

$$\mathbf{y}(n) = \mathbf{A}\mathbf{u}, \tag{5.2}$$

where

$$\mathbf{u} = [u(n), u(n-1), ..., u(n-N); u^2(n), u(n)u(n-1), ..., u(n)u(n-N)$$
$$; ...; u(n-N)u(n), ..., u^2(n-N)]^T, \tag{5.3}$$

$$\mathbf{A} = [a(0), a(1), ..., a(N), a(0,0), ..., a(0,N); ...; a(N,0), ..., a(N,N)]^T. \tag{5.4}$$

R. Pupeikis and K. Kazlauskas, *Advanced Digital Signal Processing Methods for Filtering, Identification, and Nonlinear Systems Control*, Studies in Systems, Decision and Control 638, https://doi.org/10.1007/978-3-032-06740-1_5

The output is linear and quadratic combinations of the input. The coefficients of the filter form a vector **A**. The p-th order Volterra filter implementation requires N^p multiplications for the formation of the input vector, and $O(N^p)$ computations for linear filtering. The nonlinear processor forms the input vector **u**. The input autocorrelation matrix has a large eigenvalue spread and the convergence of such linear mean square adaptive filters is slow [2]. For the slow convergence reason, Volterra filters have not been extensively applied.

The most important requirements for VLSI design are low computational load, local connectivity, and modularity. Although Volterra filters are implemented as linear filters, the obtained structures are not attractive for the VLSI implementation. The input length increases with an increase of the filter order. The adaptation is global and has to be connected to all inputs and all weights. The problem of connectivity can not be solved by systolic implementation.

The Volterra filter can also be implemented as a multichannel MA linear filters. [3] used a nonlinear combination of the input. A second order Volterra filter can be represented as follows:

$$y(n) = \sum_{i_1=0}^{N} a(i_1)u(n-i_1) + \sum_{i_2=0}^{N-1} \sum_{i_1=i_2}^{N+i_2-1} a(i_1, i_2)v(n-i_1, i_2), \tag{5.5}$$

where

$$v(n-i_1, i_2) = u(n-i_1)u(n-i_1-i_2). \tag{5.6}$$

The second order Volterra filter can be implemented as the sum of the outputs of $N+1$ linear filters. A nonlinear filter is implemented by a linear filter and nonlinear processor. A variety of algorithms have been developed based on this approach. However, lattice structures based on the multichannel approach give over-parametrized model.

Other implementation of the second order Volterra filter has been presented using a matrix representation for the second order term [1]. The second order filter output can be written as

$$\mathbf{y}(n) = \mathbf{Y}^T(n)\mathbf{V}\mathbf{Y}(n). \tag{5.7}$$

The matrix **V** can be decomposed in different ways. The matrix **V** is decomposed into a sum of matrices as follows:

$$\mathbf{y}(n) = \sum_{i=1}^{N} \varepsilon_i [\mathbf{Y}^T(n)\mathbf{Q}_i][\mathbf{Q}_i^T \mathbf{Y}(n)]. \tag{5.8}$$

The second order filter can be realized by using N linear filters each with the output $\mathbf{Q}_i^T \mathbf{Y}(n)$. Other realization are achieved by coupling the matrix decomposition approach with distributed arithmetic. These realizations are efficient only for second order filters.

Nonlinearity can also be modelled by introducing static nonlinearities at the output of linear filters. Depending on the type of nonlinearity, the model can approximate a nonlinear dynamic by realizing a Volterra series. Neural networks are examples of such implementations where a linear combiner is followed by a nonlinearity. This type of filter structure is restricted in degrees of freedom and cannot implement an arbitrary nonlinear function.

The best advantage of polynomial filters is their systematic approach toward arbitrary nonlinearities. The higher order terms can be included in the model according to the required accuracy. Another advantage is their generalized structure which includes linear filters, making polynomial filters a natural extension of linear filters. The disadvantages of these filters are overparametrization and difficult implementation procedures. The aim of this chapter is to remove these negatives by finding a computationally efficient implementation.

5.2 Layered Volterra Filter

The output $y(n)$ of any causal, discrete-time, time-invariant pth order N length nonlinear filter can be represented as a function of the input sequence $u(n)$ [6].

$$y(n) = a^{'}(0) + \sum_{i_1=0}^{N-1} a(i_1)u(n-i_1) + \sum_{i_1=0}^{N-1}\sum_{i_2=0}^{N-1} a(i_1,i_2)u(n-i_1)u(n-i_2)$$

$$+\cdots+\sum_{i_1=0}^{N-1}\ldots\sum_{i_p=0}^{N-1} a(i_1,\ldots,i_p)u(n-i_1)\ldots u(n-i_p). \tag{5.9}$$

Rearrange (5.9) as follows:

$$y(n) = a^{'}(0) + \sum_{i_1=0}^{N-1}\{a(i_1) + \sum_{i_2=0}^{N-1}\{a(i_1,i_2) + \sum_{i_3=0}^{N-1}\{a(i_1,i_2,i_3)+\cdots$$

$$+\sum_{i_{p-1}=0}^{N-1}\{a(i_1,\ldots,i_{p-1}) + \sum_{i_p=0}^{N-1} a(i_1,\ldots,i_p)u(n-i_p)\}u(n-i_{p-1})\}\ldots\}u(n-i_2)\}u(n-i_1). \tag{5.10}$$

From (5.10) we obtain a layered representation of the p-th order Volterra filter (5.9):

1-st layer: $y(n) = a^{'}(0) + \sum_{i_1=0}^{N-1} \alpha_2(n,i_1)u(n-i_1)$,

2-nd layer: $\alpha_2(n,i_1) = a(i_1) + \sum_{i_2=0}^{N-1} \alpha_3(n,i_1,i_2)u(n-i_2)$,

$$\ldots \tag{5.11}$$

$(p-1)$-st layer: $\alpha_{p-1}(n, i_1, ..., i_{p-2}) = a(i_1, ..., i_{p-2}) + \sum_{i_{p-1}=0}^{N-1} \alpha_p(n, i_1, ..., i_{p-1})$ $u(n - i_{p-1})$,

p-st layer: $\alpha_p(n, i_1, ..., i_{p-1}) = a(i_1, ..., i_{p-1}) + \sum_{i_p=0}^{N-1} a(i_1, ..., i_p)u(n - i_p)$.

The pth order Volterra filter is realized with p layers of MA filters. The 1st layer is an MA filter with variable coefficients. The coefficients of the 1st layer are modelled with N MA filters of the 2nd layer. The coefficients of the 2nd layer are modelled with N^2 MA filters of the 3rd layer, and so on. The pth layer has N^{p-1} linear MA filters with constant coefficients. The p-th order of nonlinearity is described by the $(p-1)$-th layer variations.

For example, if p=3, then the 3rd order Volterra filter has three layers:

1st layer: $y(n) = \acute{a}(0) + \sum_{i_1=0}^{N-1} \alpha_2(n, i_1)u(n - i_1)$,

2nd layer: $\alpha_2(n, i_1) = a(i_1) + \sum_{i_2=0}^{N-1} \alpha_3(n, i_1, i_2)u(n - i_2)$,

3nd layer: $\alpha_3(n, i_1, i_2) = a(i_1, i_2) + \sum_{i_3=0}^{N-1} a_3(i_1, i_2, i_3)u(n - i_3)$.

The 3-rd order layered nonlinear Volterra filter structure is shown in Fig. 5.1.

5.3 Layered PAR Filter

PAR filter can be implemented in a layered manner using linear AR filters. Any pth order N length PAR filter can be implemented in p layers. The jth layer consists of N^{j-1} AR filters. The pth order N length PAR filter is

$$y(n) = u(n) + a(0) + \sum_{i_1=1}^{N} a(i_1)y(n - i_1) + \sum_{i_1=1}^{N}\sum_{i_2=1}^{N} a(i_1, i_2)y(n - i_1)y(n - i_2)$$

$$+ ... + \sum_{i_1=1}^{N} \cdots \sum_{i_p=1}^{N} a(i_1, ..., i_p)y(n - i_1)...y(n - i_p). \tag{5.12}$$

We have from (5.12)

$$y(n) = u(n) + a(0) + \sum_{i_1=1}^{N} \Bigg\{ a(i_1) + \sum_{i_2=1}^{N} \Bigg\{ a(i_1, i_2) + \cdots + \sum_{i_{p-2}=1}^{N} \Bigg\{ a(i_1, ..., i_{p-2})$$

$$+ \sum_{i_{p-1}=1}^{N} \Bigg\{ a(i_1, ..., i_{p-1}) + \sum_{i_p=1}^{N} a(i_1, ..., i_p)y(n - i_p) \Bigg\} y(n - i_{p-1}) \Bigg\} ... \Bigg\} y(n - i_2) \Bigg\} y(n - i_1). \tag{5.13}$$

Then, from (5.13) we obtain the following layered representation of the p-th order PAR filter (5.12):

1st layer: $y(n) = u(n) + a(0) + \sum_{i_1=1}^{N} \beta_2(n, i_1)y(n - i_1)$,

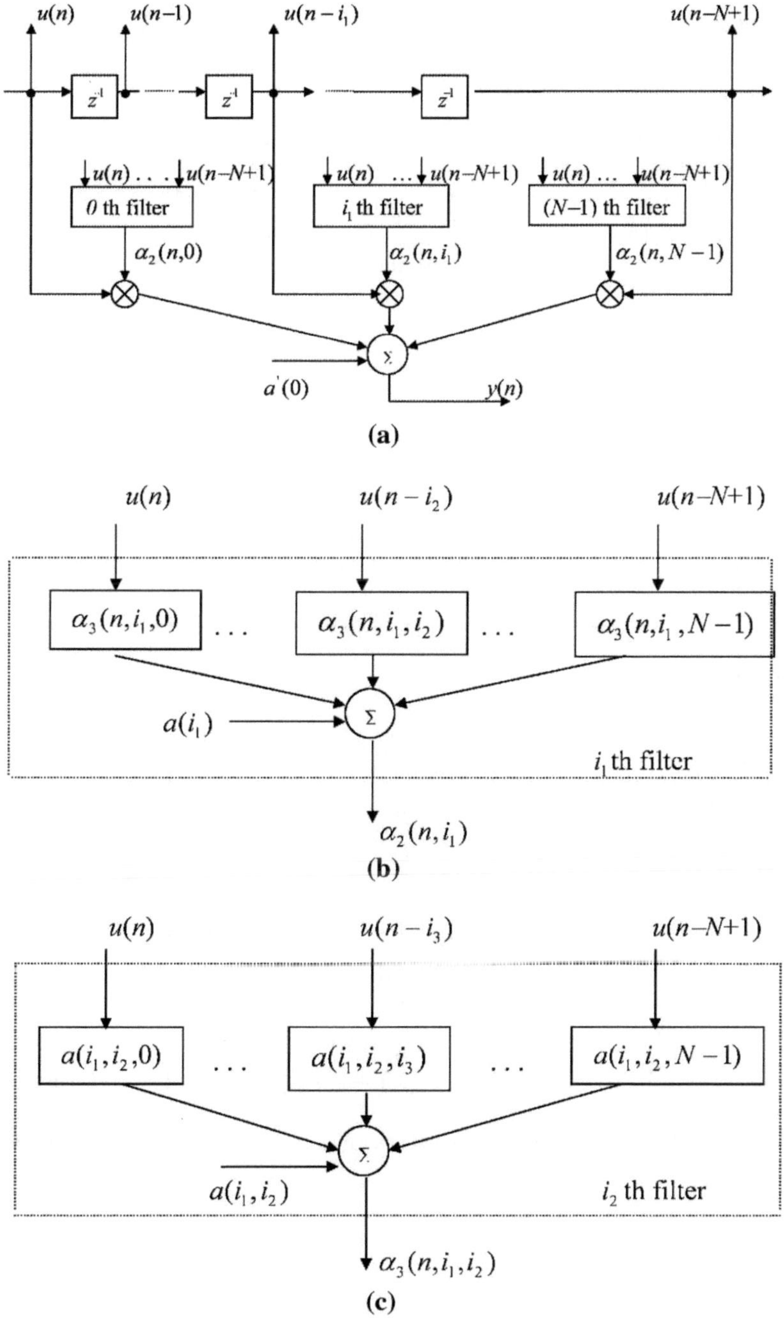

Fig. 5.1 The 3rd order layered nonlinear Volterra filter structure. **a** 1st layer filter, **b** i_1th filter of the 2nd layer, **c** i_2th filter of the 3rd layer.*Note*: in section (**a**) a small rectangle marks the delay unit z^{-1}

2nd layer: $\beta_2(n, i_1) = a(i_1) + \sum_{i_2=1}^{N-1} \beta_3(n, i_1, i_2)y(n - i_2), \cdots\cdots\cdots$

$(p-1)$-st layer: $\beta_{p-1}(n, i_1, ..., i_{p-2}) = a(i_1, ..., i_{p-2}) + \sum_{i_{p-1}=1}^{N} \beta_p(n, i_1, ..., i_{p-1})y(n - i_{p-1})$,

p-th layer: $\beta_p(n, i_1, ..., i_{p-1}) = a(i_1, ..., i_{p-1}) + \sum_{i_p=1}^{N} a(i_1, ..., i_p)y(n - i_p)$.

As an example, a second order PAR filter has two layers:

1st layer: $y(n) = u(n) + a(0) + \sum_{i_1=1}^{N} \beta_2(n, i_1)y(n - i_1)$,

2nd layer: $\beta_2(n, i_1) = a(i_1) + \sum_{i_2=1}^{N-1} a(i_1, i_2)y(n - i_2)$,

The 2nd order PAR filter structure is shown in Figs. 5.2 and 5.3.

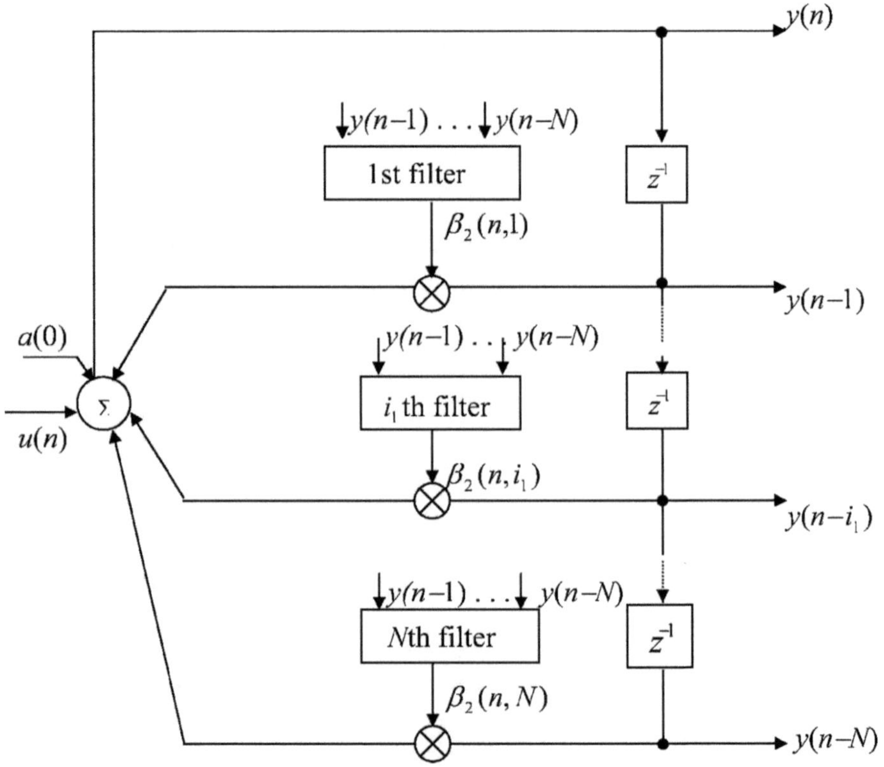

Fig. 5.2 The 2nd order layered PAR filter: 1st layer filter. *Note*: a small rectangle marks the delay unit z^{-1}

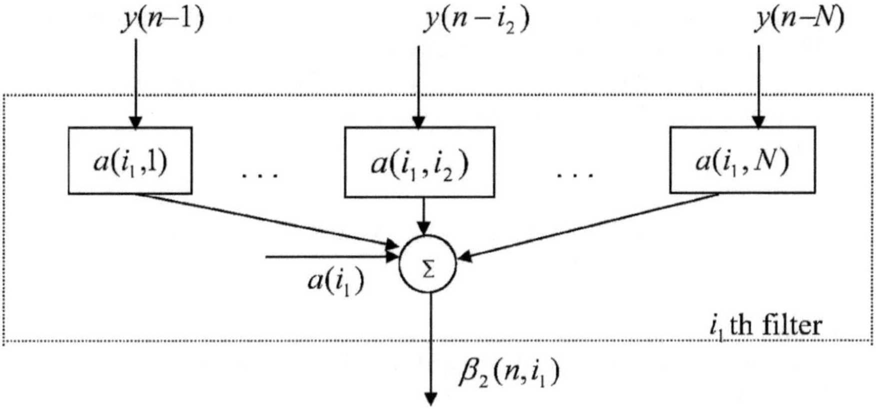

Fig. 5.3 The 2nd order layered PAR filter: i_1-th filter of the 2nd layer

5.4 Layered Bilinear Filter

The discrete-time bilinear filter has the form [5]:

$$y(n) = a(0) + \sum_{i_1=1}^{N} a(i_1)y(n-i_1) + \sum_{i_2=0}^{N-1} c(i_2)u(n-i_2)$$

$$+ \sum_{i_1=1}^{N} \sum_{i_2=0}^{N-1} b(i_1,i_2)y(n-i_1)u(n-i_2). \tag{5.14}$$

We obtain from (5.14)

$$y(n) = a(0) + \sum_{i_1=1}^{N} a(i_1)y(n-i_1) + \sum_{i_2=0}^{N-1} \left\{ c(i_2) + \sum_{i_1=1}^{N} b(i_1,i_2)y(n-i_1) \right\} u(n-i_2). \tag{5.15}$$

and

$$y(n) = a(0) + \sum_{i_2=0}^{N-1} c(i_2)u(n-i_2) + \sum_{i_1=1}^{N} \left\{ a(i_1) + \sum_{i_2=0}^{N-1} b(i_1,i_2)u(n-i_2) \right\} y(n-i_1). \tag{5.16}$$

Then, we have from (5.15) the First layered bilinear filter:
1st layer: $y(n) = a(0) + \sum_{i_1=1}^{N} a(i_1)y(n-i_1) + \sum_{i_2=0}^{N-1} \gamma(n,i_2)u(n-i_2)$,
2nd layer:

$$\gamma(n,i_2) = c(i_2) + \sum_{i_1=0}^{N} b(i_1,i_2)y(n-i_2). \tag{5.17}$$

The AR part has a linear dependence, and the MA part has a nonlinear dependence. The MA part has output dependent coefficients.

The First layered bilinear filter structure is shown in Figs. 5.4 and 5.5.

We obtain from (5.16) the Second layered bilinear filter:

1st layer:

$$y(n) = a(0) + \sum_{i_2=0}^{N-1} c(i_2)u(n-i_2) + \sum_{i_1=1}^{N-1} \delta(n, i_1)y(n-i_1), \tag{5.18}$$

2nd layer: $\delta(n, i_1) = a(i_1) + \sum_{i_2=0}^{N-1} b(i_1, i_2)u(n-i_2)$.

The MA part has a linear dependence, and the AR part has a nonlinear dependence. The AR part has input dependent coefficients.

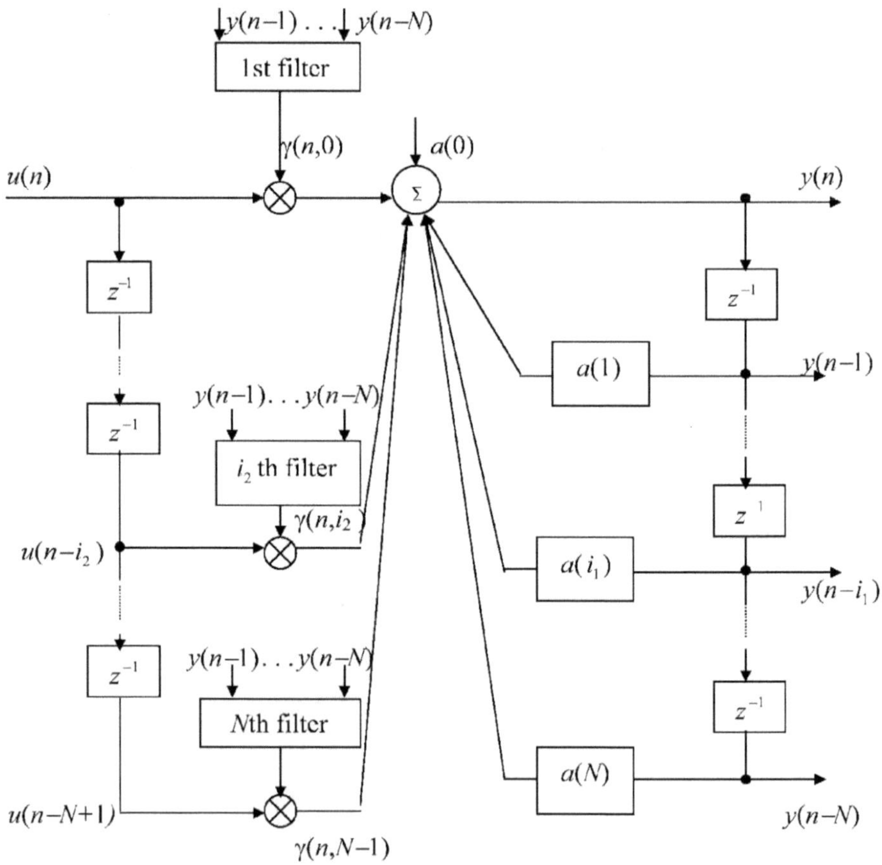

Fig. 5.4 The First layered bilinear filter structure: 1st layer. *Note*: a small rectangle marks the delay unit z^{-1}

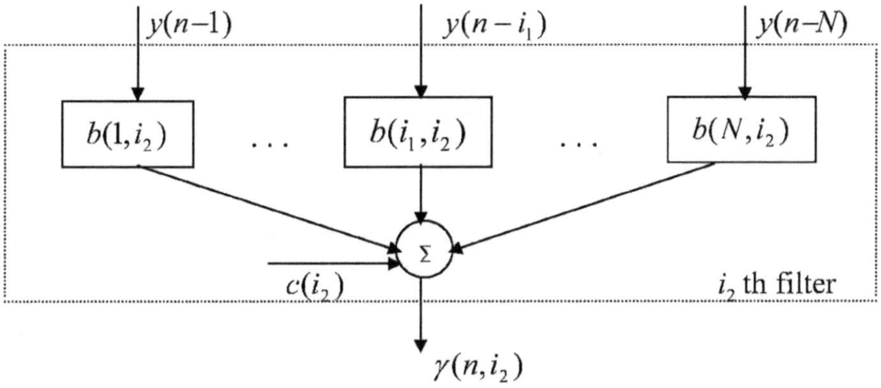

Fig. 5.5 The First layered bilinear filter structure: i_2th filter of the 2nd layer

The Second layered bilinear filter flowchart is shown in Figs. 5.6 and 5.7.

The number of multiplications per time instant for the bilinear filter (5.14) is $2N(N+1)$. The number of multiplications per time instant for the First layered bilinear filter and for Second layered bilinear filter is $N(N+2)$.

Table 5.1 shows the number of multiplications per one time instant for bilinear filter (5.14) and for the Second layered bilinear filters (5.17) and (5.18).

5.5 Stability

Variable nature of the layered filter helps us define a dynamical stability for such filters. Stability of the usual filters depends on the location of poles and zeros in the z plane. We extend it to nonlinear filters using their layered descriptions. Due to the filters with variable parameters in the layered structure, the location of the poles and zeros also vary with time. Therefore, we call such filters to have variable poles and zeros. From this definition we can argue that nonrecursive filters with variable parameters are stable in the sense that, as long as their parameter values are finite, they satisfy BIBO stability. Each module in the layered structure of such filters will have a bounded output and thus the variable parameter of the above module will remain finite and result in overall nonlinear filter stability.

A recursive filter with variable parameters and its stability analysis is more complicated. A stability can be enforced by limiting the output of the second layer filters such that the variable coefficients of the 1-st layer for all instants remain within the unit circle on the z plane. A more weak condition can be defined by allowing the variable poles for finite time to go outside the unit circle. During their visit to the unstable zone, the variable poles must have an effect to on the overall system states such that they diminish in magnitude thus returning within the unit circle.

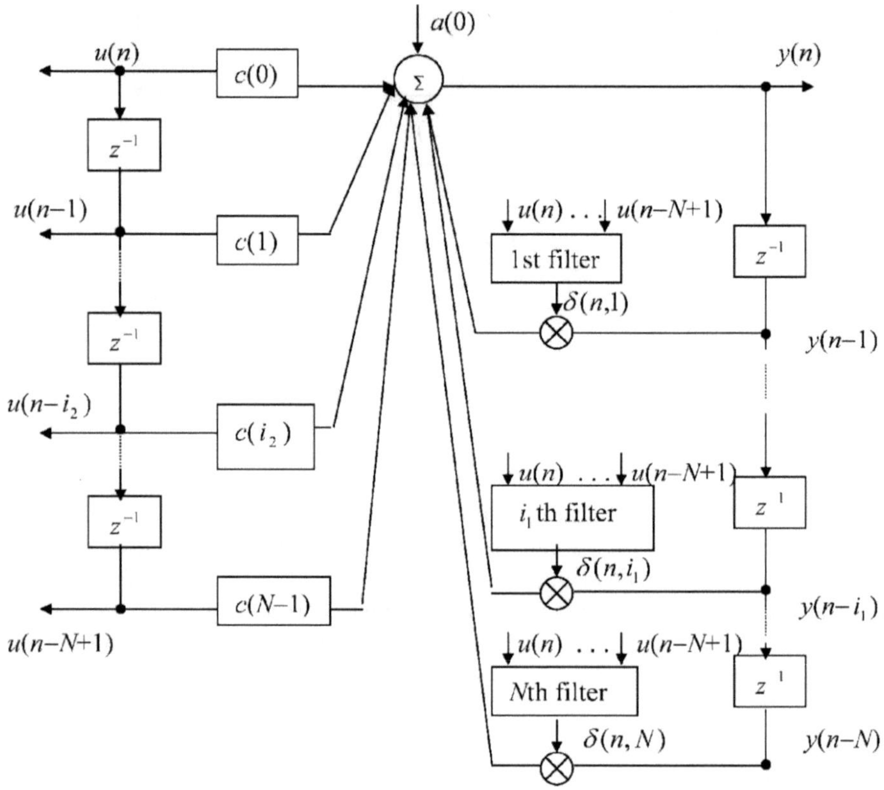

Fig. 5.6 The Second layered bilinear filter structure: 1st layer. *Note*: a small rectangle marks the delay unit z^{-1}

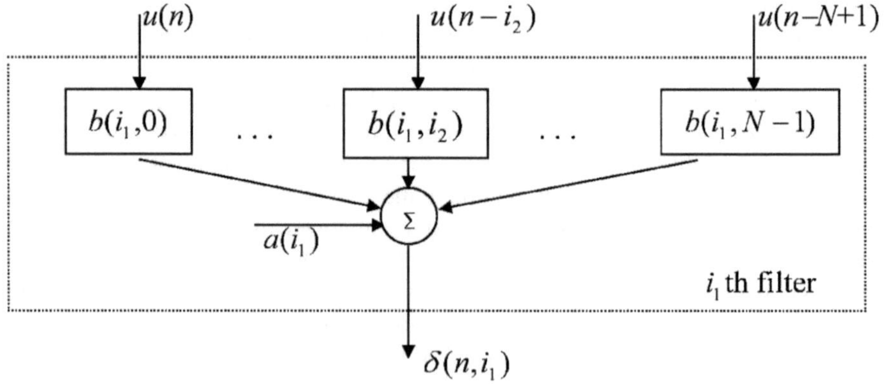

Fig. 5.7 The Second layered bilinear filter structure: i_1th-filter of the 2nd layer

Table 5.1 Multiplication complexity comparison for bilinear and layered bilinear filters

Filter length N	Bilinear filter (5.14)	Layered bilinear filters (5.15) and (5.16)
10	220	120
20	840	440
30	1860	960
40	3280	1680
50	5100	2600
100	20200	10200

5.6 Conclusion

The polynomial nonlinear filters can be efficiently implemented through layered structures. The layered structures of polynomial nonlinear filters are computationally efficient, have a modular structure, and can be implemented using simple MA or AR linear filters.

References

1. Chiang, H., Nikias, C.L., Venetsanopoulos, A.N.: Efficient implementation of quadratic digital filters. IEEE Trans. ASSP **34**(6), 1511–1528 (1986)
2. Mathews, V.J., Lee, J.: A fast recursive least squares second order Volterra filter. Proc. ICASSP 1383–1386 (1988)
3. Perry, F.A., Parker.: Adaptive solution of multichannel lattice models for linear and nonlinear systems. Proc. ICASSP 744–746 (1980)
4. Priestley, M.B.: State dependent models: a general approach to time series analysis. J. Time Ser. Anal. **1**(1), 47–71 (1980)
5. Priestley, M.B.: Non-Linear and Non-Stationary Time Series Analysis. Academic (1998)
6. Raz, G.M., Van Veen, B.D.: Baseband Volterra filters for implementing carier based nonlinearities. IEEE Trans. Signal Proc. **6**(1), 103–114 (1998)

Chapter 6
Exploring the Potentiality of Nonlinear Systems for Minimum Variance Control

The objective of the chapter is to explore potentials for adaptive MV control of ARMAX systems having PWL nonlinearities. A basic introduction to a digital analysis of linear and nonlinear ARMAX systems is given. Two types of invertible nonlinearity system models, the Wiener system and the Hammerstein one, are analysed in a closed-loop framework. For both systems a good many examples, based numerical simulation, are given.

6.1 State-Space Equations of a Stochastic System

Consider a stochastic system

$$\mathbf{x}(k+1) = \mathbf{\Phi}\mathbf{x}(k) + \mathbf{\Psi}\mathbf{u}(k) + \mathbf{\Gamma}\mathbf{w}(k), \tag{6.1}$$

$$y(k) = \mathbf{H}f(\mathbf{x}(k);\eta) + v(k), \tag{6.2}$$

including the known static invertible nonlinearity $f(\mathbf{x}(k);\eta)$, where $\mathbf{x}(k) \in \mathbf{R}^n$ is a state, $\mathbf{u}(k) \in \mathbf{R}^m$, $\mathbf{z}(k) \in \mathbf{R}^p$ and $\mathbf{y}(k) \in \mathbf{R}^r$ are input, noisy intermediate output, and noisy measurable output of the system, respectively, $\mathbf{w}(k)$ and $v(k)$ are zero-mean disturbance terms with covariance matrices $\mathbf{Q}$, and $\mathbf{G}$, respectively, and $\mathbf{\Phi}$, $\mathbf{\Psi}$, $\mathbf{\Gamma}$ and $\mathbf{H}$ are the known system matrices [14], k is the current number of observation samples.

Assume that the static nonlinearity $f(\mathbf{x}(k);\eta)$ $\forall$ k=1,2,..., N is absent.

Then Eq. (6.2) turns into

$$y(k) = \mathbf{H}\mathbf{x}(k) + v(k). \tag{6.3}$$

R. Pupeikis and K. Kazlauskas, *Advanced Digital Signal Processing Methods for Filtering, Identification, and Nonlinear Systems Control*, Studies in Systems, Decision and Control 638, https://doi.org/10.1007/978-3-032-06740-1_6

Introducing a single stage performance index we have (Mišković et al. [14]):

$$J(u) = \mathcal{E}\{\mathcal{L}[x(k+1), u(k)]\}, \tag{6.4}$$

$$\mathcal{L}[x(k+1), u(k)] = \begin{bmatrix} x(k+1) \\ u(k) \end{bmatrix}^T \begin{bmatrix} \mathbf{W} & \mathbf{S} \\ \mathbf{S}^T & \mathbf{U} \end{bmatrix} \begin{bmatrix} x(k+1) \\ u(k) \end{bmatrix}, \tag{6.5}$$

where **W** and **S** are constant, bounded and non-negative definite weighting matrices ($\mathbf{W} \geq 0$, $\mathbf{S} \geq 0$), while **U** is a constant, bounded and positive definite weighting matrix ($\mathbf{U} > \mathbf{0}$) and $\mathcal{L}$ is a monic prefilter, possibly dependent on the parameters, that can be used to enhance certain frequency regions, and $\mathcal{E}$ is a mathematical expectation.

The control, obtained by minimizing (6.4), (6.5), will be referred as the MSEV control and is given by Mišković et al. [14]:

$$\mathbf{u}(k) = \mathbf{M}\hat{\mathbf{x}}(k);\ \hat{\mathbf{x}}(k) = \mathcal{E}\{\mathbf{x}(k)|\mathbf{Y}(k-1)\}, \tag{6.6}$$

$$\mathbf{M} = -(\mathbf{\Psi}^T\mathbf{W}\mathbf{\Psi} + \mathbf{U} + 2\mathbf{\Psi}^T\mathbf{S})^{-1}(\mathbf{\Psi}\mathbf{W}\mathbf{\Phi} + \mathbf{S}^T\mathbf{\Phi}). \tag{6.7}$$

Here, $\mathbf{Y}(k-1) = (y(0), y(1), ..., y(k-1))$ is the set of measurements up to time $(k-1)T_s$, where T_s is the sampling period, chosen according to the Nyquist theorem, $\hat{\mathbf{x}}(k)$ represents the minimum variance one-step-ahead prediction of the state $\mathbf{x}(k)$ and may be generated using the well-known Kalman predictor, which requires the noise statistics **Q** and **G** to be known exactly [1, 3].

Usually the parameters of system (6.1)–(6.3) have to be estimated just before the control algorithm can be implemented. To ensure identifiability, model (6.1)–(6.3) is replaced by the steady-state innovations model in observer canonical form [2, 13]:

$$\mathbf{x}(k+1) = \mathbf{\Phi}\mathbf{x}(k) + \mathbf{\Psi}\mathbf{u}(k) + \mathbf{\Gamma}e(k), \tag{6.8}$$

$$y(k) = \mathbf{H}\mathbf{x}(k) + e(k), \tag{6.9}$$

where matrices $\mathbf{\Psi}$ and H are as follows:

$$\mathbf{\Psi} = \left[\begin{array}{c|ccc} -a_1 & & & \\ -a_2 & & & \\ \cdot & & & \\ \cdot & & \mathbf{I}_{(n-1)\times(n-1)} & \\ \cdot & & & \\ -a_{n_a-1} & -- & -- & -- \\ -a_{n_a} & 0 & \dots & 0 \end{array}\right]; \tag{6.10}$$

$$\mathbf{H}^T = \begin{bmatrix} 1 \\ \cdot \\ \cdot \\ \cdots \\ 0 \end{bmatrix}; \mathbf{\Phi} = \begin{bmatrix} \phi_1 \\ \cdot \\ \cdot \\ \cdots \\ \phi_n \end{bmatrix}; \mathbf{\Psi} = \begin{bmatrix} \psi_1 \\ \cdot \\ \cdot \\ \cdots \\ \psi_n \end{bmatrix}; \mathbf{\Gamma} = \begin{bmatrix} \gamma_1 \\ \cdot \\ \cdot \\ \cdots \\ \gamma_n \end{bmatrix} \tag{6.11}$$

with $\mathbf{I}_{(n-1)\times(n-1)}$ being the identity matrix of the order $n-1$.

6.2 ARMAX Model Parameter Estimation

The problem of parameter estimation of the observer state space form (6.8)–(6.11) can be solved by taking advantage of the fact that the model may be equivalently expressed in the form of a n-dimensional ARMAX model with external input according to Anderson and Moore [2], Goodwin and Sin [6], Miškovič et al. [14]:

$$A(q^{-1})y(k) = B(q^{-1})u(k) + C(q^{-1})e(k), \tag{6.12}$$

where q^{-1} is the backward time-shift operator such that $q^{-1}u(k) = u(k-1)$. The polynomials $A(q^{-1})$, $B(q^{-1})$, and $C(q^{-1})$ are:

$$A(q^{-1}) = 1 + \sum_{i=1}^{n_a} a_i q^{-i},\ B(q^{-1}) = 1 + \sum_{i=1}^{n_b} b_i q^{-i},\ C(q^{-1}) = 1 + \sum_{i=1}^{n_c} c_i q^{-i}, \tag{6.13}$$

where n_a, n_b and n_c are orders of the respective polynomials.

Coefficients $a_i\ \forall i = \overline{1, n_a}$, $b_i \forall i = \overline{1, n_b}$, and $c_i\ \forall i = \overline{1, n_c}$ are obtained from elements of matrices $\mathbf{\Psi}$ and $\mathbf{\Gamma}$ [13, 14]:

$$\psi = b_i,\ \gamma_i = c_i - a_i,\ i = 1, ..., n. \tag{6.14}$$

Moreover, it is possible to express ARMAX (6.12) in the linear regression form as follows:

$$y(k) = \mathbf{Z}^T(k)\mathbf{\Theta} + e(k), \tag{6.15}$$

where $\mathbf{\Theta}$ is the vector of unknown parameters given as

$$\mathbf{\Theta} = (a_1, ..., a_{n_a}, b_1, ..., b_{n_b}, c_1, ..., c_{n_c})^T \tag{6.16}$$

and $\mathbf{Z}(k)$ is a regression vector containing the appropriate set of past outputs, inputs, and innovations [14]:

$$\mathbf{Z}(k) = [-y(k-1), ..., -y(k-n_{n_a}), u(k-1), ..., u(k-n_{n_b}),$$

$$e(k-1), ..., e(k-n_{n_c})]^T. \tag{6.17}$$

Once the regression model (6.15)–(6.17) is obtained, a recursive estimate of the parameter vector $\mathbf{\Theta}$ may be achieved using a number of alternative algorithms, presented in Ljung and Söderström [13], Goodwin and Sin [6]. The ordinary parametric identification algorithms require that:

- the input signal should assure condition of the persistently excitation of the availaible system;
- the static PWL nonlinearity should be invertible;
- the linear dynamic block should be of the *minimum phase*;
- the over-parameterization problem while parameter estimates calculation should be avoided;
- the intensity of the process noise $w(k)$ and measurement noise $v(k)$ should be small, as far as possible.
- the additive measurement noise filter should be of special form.

The PLR algorithm can be written as follows Ljung and Gunnarsson [12]:

$$\hat{\mathbf{\Theta}}(k) = \hat{\mathbf{\Theta}}(k-1) + \mathbf{\Phi}(k)\epsilon(k), \tag{6.18}$$

$$\epsilon(k) = y(k) - \mathbf{Z}^T(k)\hat{\mathbf{\Theta}}(k-1), \tag{6.19}$$

$$\mathbf{\Phi}(k) = \mathbf{\Gamma}(k-1)\mathbf{Z}(k)[\beta(k) + \mathbf{Z}^T(k)\mathbf{\Gamma}(k-1)\mathbf{Z}(k)]^{-1}, \tag{6.20}$$

$$\mathbf{\Gamma}(k) = \beta^{-1}(k)\{\mathbf{\Gamma}(k-1) - \frac{\mathbf{\Gamma}(k-1)\mathbf{Z}(k)\mathbf{Z}^T(k)\mathbf{\Gamma}(k-1)}{\beta(k) + \mathbf{Z}^T(k)\mathbf{\Gamma}(k-1)\mathbf{Z}(k)}\}. \tag{6.21}$$

Here $\hat{\mathbf{\Theta}}$ is unknown parameter estimates vector to be calculated by PLR procedure, besides $\hat{\mathbf{\Theta}}^T=(\hat{\mathbf{a}}^T, \hat{\mathbf{b}}^T, \hat{\mathbf{c}}^T)$, $\hat{\mathbf{a}} = (\hat{a}_1, ..., \hat{a}_{n_a})^T$, $\hat{\mathbf{b}} = (\hat{b}_1, ..., \hat{b}_{n_b})^T$, $\hat{\mathbf{c}} = (\hat{c}_1, ..., \hat{c}_{n_c})^T$; $0 < \beta(k) \leq 1$ is the time-varying factor introduced to forget a previous information.

These recursive equations are initialized with an assumed initial estimate $\hat{\mathbf{\Theta}}(0)$ and initial covariance matrix $\mathbf{\Gamma}(0)$. Under certain conditions estimates may be expected to converge to the correct parameter values [6, 8, 13].

6.3 Controller Design for ARMAX System

Suppose that an open-loop LTI control system to be observed is SISO, i.e. with one output $y(k)$ and one input $u(k)$, represented by the ARMAX model

$$y(k) = \underbrace{G_0(q^{-1}; \mathbf{\Theta})q^{-\tau}u(k)}_{x(k)} + \underbrace{\lambda H_0(q^{-1}; \varphi)\xi(k)}_{v(k)}, \tag{6.22}$$

that consists of two parts: an open system transfer function

$$G_0(q^{-1};\Theta) = \frac{B(q^{-1},\mathbf{b})}{A(q^{-1},\mathbf{a})}, \tag{6.23}$$

and a measurement noise $v(k) \sim \mathcal{N}(0, \sigma_v)$, where

$$H_0(q^{-1};\varphi) = \frac{C(q^{-1},\mathbf{c})}{A(q^{-1},\mathbf{a})} \tag{6.24}$$

is an inversely stable monic filter used to generate a noise.

Here λ is the positive multiplier; τ is the integer that corresponds to the pure time delay, which is the respective samples number multiplier; $\xi(k)$ is used to generate the unmeasurable additive measurement noise $v(k)$ [5], applying the filter $H_0(q^{-1},\varphi)$, $\xi(k)$ is independent statistically and stationary with the properties

$$\mathcal{E}\{\xi(k)\} = 0,\ \mathcal{E}\{\xi(k)\xi(k+\tau)\} = \sigma_\xi^2\delta(\tau), \tag{6.25}$$

σ_ξ^2 is the variance of measurement noise sequence $\xi(k)$, $\delta(\tau)$ is the Kronecker delta function.

The flowchart of the LTI system with the feedback is represented in Fig. 6.1. Here the current sample of an input signal $u(k)$ is expressed by

$$u(k) = \underbrace{[r(k) - y(k)]}_{e(k)} G_R(q^{-1};\alpha), \tag{6.26}$$

where $e(k)$ is a prediction error between current samples of a set-point and output signals, α is the parameter vector of the controller $G_R(q^{-1};\alpha)$, $r(k)$ is the current sample of a reference (set-point) signal $r(k)$, which is a quasi-stationary signal, independent of the stochastic disturbance $v(k)$.

The LTI controller $G_R(q,\alpha)$ is build by minimizing the exponentially stable quadratic performance function

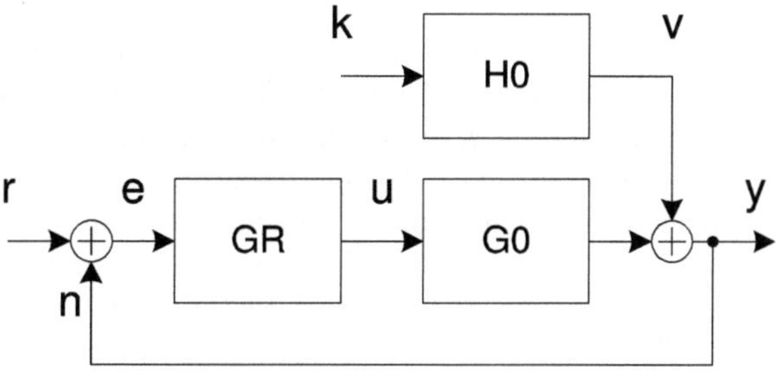

Fig. 6.1 A LTI system acting in a closed-loop

$$J = \lim_{N\to\infty} \mathcal{E}\{\frac{1}{N}\sum_{k=0}^{N}\{[py(k) - gr(k)]^2 + (qu)^2(k)\}\}. \quad (6.27)$$

Here p, g and q are weighting factors; N is a total number of observations to be processed.

In Fig. 6.1: $G_R \equiv G_R(q^{-1}, \alpha)$ is a ST MV controller; $G_0 \equiv G_0(q^{-1}, \Theta(k))$ is the LTI system, and $H_0 \equiv H_0(q^{-1}, \varphi)$ is a filter of an additive noise, α, $\Theta(k)$, φ are vectors of unknown parameters. The signals are as follows: $u(k)$ is an controller's output, $y(k)$ is an output corrupted by noise, $v(k)$ is a measurement noise, $\zeta(k)$ is a sequence of independent Gaussian variables, $e(k)$ characterizes how closely the output follows a set-point signal in a closed-loop flowchart.

The general form of the MV control law is

$$u(k) = \frac{p(\sum_{i=1}^{n_a} a_i y(k+\tau-i) - \sum_{i=1}^{n_b} b_i u(k-i)) + gr(k)}{p^2 b_0^2 + q}, \quad (6.28)$$

It is obtained by minimizing the performance function

$$J_{GMV} = \mathcal{E}\{[(py(n_a+\tau) - gr(k))^2 + q'^2 u^2(n_b)] \,|_t\}, \quad (6.29)$$

that rewritten in an expanded form according to Evans 2008 is

$$\tilde{J}_{GMV} = p^2[u^2(k)b_0^2 + 2u(t)b_0 u(k-i)\sum_{i=1}^{n_b} b_i -$$

$$- 2u(t)b_0 y(k+\tau-i)\sum_{i=1}^{n_a} a_i - 2u(k)b_0 gr(k)] + q'^2 u^2(k). \quad (6.30)$$

Here $n_b \geq n_a$; p, g, q and q' are scalar coefficients, besides

$$q = \frac{q'^2}{pb_0} \quad (6.31)$$

in accordance to Evans [7] and Tham [15], t is a current time.

To implement the self-tuning MV controller, it is necessary in each current recursive operation, firstly, to estimate system's linear block parameters in a noisy environment and, secondly, to determine the controller coefficients that are recalculated using the values of abovementioned estimates. Thus, at each current time moment kT_s $\forall k = 1, 2,...$ it is needed to find the estimate of the parameter vector $\mathbf{\Theta}(k)$ in order to change in Eq.(6.28) the true but unknown coefficients $b_0(k), b_1(k), \ldots, b_{n_{b(k)}}, a_1(k), \ldots, a_{n_{a(k)}}$ by their current estimates $\hat{b}_0(k), \hat{b}_1(k), \ldots,$ $\hat{b}_{n_{b(k)}}, \hat{a}_1(k), \ldots, \hat{a}_{n_{a(k)}}$, i.e.:

$$u(k) = \frac{p(\sum_{i=1}^{n_a} \hat{a}_i(k)y(n_a + k - i) - \sum_{i=1}^{n_b} \hat{b}_i(k)u(n_b + k - i)) + gr(k)}{p\hat{b}_0(k) + q}, \tag{6.32}$$

respectively.

6.4 Block-Oriented Wiener and Hammerstein Systems

Suppose that checking the system for linearity we are noticed - the digital system with the output

$$y(k) = G_0(q^{-1}; \Theta(k))u(k) + \lambda H_0(q^{-1}; \varphi)\xi(k) \tag{6.33}$$

is not the LTI system. Note that for simplicity in (6.33) multiplier $\lambda = 1$. We guess that it could be a nonlinear system. The true output of the block-oriented nonlinear system can be expressed as follows

$$y_w(k) = f\{G_0(q^{-1}; \Theta(k))u(k) + w(k); \eta\} + H_0(q^{-1}; \varphi)\xi(k), \tag{6.34}$$

if a linear block follows a static nonlinearity and

$$y_h(k) = [f\{(u(k), \eta)\} + w(k)]\{G_0(q^{-1}; \boldsymbol{\Theta}(k)) + v(k)\} \tag{6.35}$$

if a static nonlinearity precedes a linear block. Here indexes w and h denote names of Wiener and Hammerstein systems.

Let us assume now, for simplicity, that $H_0(q^{-1}; \varphi) \equiv 1$, and both noises in (Figs.1.2 and 1.3) are absent, too. Then, the true form of the closed-loop system input $u(k) \equiv u_w(k)$ is

$$u_w(k) = \underbrace{[r(k) - \hat{y}_w(k)]}_{e_w(k)} G_R(q^{-1}; \alpha), \tag{6.36}$$

while the output has the form

$$\hat{y}_w(k) = f\{G_0(q^{-1}; \Theta(k))u(k), \eta\} \tag{6.37}$$

if instead of the LTI system, the Wiener system is present. The input and output are

$$u_h(k) = \underbrace{[r(k) - \hat{y}_h(k)]}_{e_h(k)} G_R(q^{-1}; \alpha), \tag{6.38}$$

$$\hat{y}_h(k) = f\{(u(k), \eta)\}G_0(q^{-1}; \boldsymbol{\Theta}(k)), \tag{6.39}$$

respectively, if instead of the LTI system the Hammerstein system is available. The LTI controller $G_R(q, \alpha)$ is built, using the function (6.27). Indeed, the performance functions

$$J_w = \lim_{N\to\infty} \mathcal{E}\{\frac{1}{N}\sum_{k=0}^{N}\{[p_w\tilde{y}_w(k) - g_w r(k)]^2 + \gamma_w u_w^2(k)\}\}, \tag{6.40}$$

$$J_h = \lim_{N\to\infty} \mathcal{E}\{\frac{1}{N}\sum_{k=0}^{N}\{[p_h\hat{y}_h(k) - g_h r(k)]^2 + \gamma_h u_h^2(k)\}\} \tag{6.41}$$

are minimized if instead of the LTI system, the Wiener or Hammerstein systems are present, respectively. Note that in (6.40):

$$\tilde{y}_w(k) \equiv \hat{y}_w(k) f^{-1}(\cdot, \eta). \tag{6.42}$$

The current values of control signals are:

$$u_w(k) = \frac{p_w(\sum_{i=1}^{n_a}\hat{a}_i(k)\tilde{y}_w(n_a + k - i) - \sum_{i=1}^{n_b}\hat{b}_i(k)u_w(n_b + k - i)) + g_w r(k)}{p_w\hat{b}_0(k) + q_w} \tag{6.43}$$

for the Wiener system, and

$$u_h(k) = \frac{p_h(\sum_{i=1}^{n_a}\hat{a}_i(k)\hat{y}_h(n_a + k - i) - \sum_{i=1}^{n_b}\hat{b}_i(k)u_h(n_b + k - i)) + g_h r(k)}{p_h\hat{b}_0(k) + q_h} \tag{6.44}$$

for the Hammerstein one. Equation (6.43) coincides with Eq. (6.32) if $p_w \equiv p$, $g_w \equiv g$, $q_w \equiv q$, and (6.42) is valid.

The MV law could be used for the Wiener system that is transformed into the system with reduced nonlinearity under Eq. (6.42). In such a case, the nonlinear filter, inverse to the Wiener system nonlinearity, is applied to reconstruct the current value of an intermediate signal recursively. It is necessary here to determine the parameter vector α of the LTI controller $G_R(q^{-1}; \alpha)$ simultaneously with the current parameter vector $\Theta(k)$ inside the permissible area of their stability. Therefore, the adaptive parametric identification, as compared with batch processing will be preferable. However, as a rule, sometimes there appeares there an unstable model because of non-accurate parameter estimates in a current iteration k. Note that they are substituted in the control law equations instead of the unknown true parameter values automatically.

6.5 Closed-Loop Parametric Identification Method

The quality of performance of an adaptive control system depends on the accuracy of a tuned process model and on the identification method to be applied [10]. It is known that parameter estimates are substituted instead of true but unknown parameters (Fig. 6.2) in the control law to be used in adaptive control system flowchart.

According to Forssell and Ljung [5] the closed-loop identification methods fall into the following groups: the direct approach, the indirect approach, and the joint input-output approach. The direct approach identifies the open-loop system using measurements of the input-output signals. The indirect approach identifies some closed-loop transfer function and determines the open-loop parameters using the knowledge of the controller. The joint input-output approach regards the input and output jointly as the output from a system driven by some extra input or set-point signal and noise.

The general flowchart of parameter adaptive self-organizing system, given by Evans [7], is shown in Fig. 6.2. Three parts are dominant in Fig. 6.2: a system part, an estimator part, and a controller one. Recursive estimation of the process model parameters is carried out in the part of estimator. It is used to predict the value $\hat{y}(k)$ of process output $y(k)$ as well. The last part utilizes the current parameter estimates $\hat{\Theta}(k)$ produced by estimator. The control part consists of a block that is built up to calculate the controller parameters using the plant model parameter estimates $\hat{\Theta}(k)$, and applying the samples of two signals: output $y(k)$ and extra-signal $r(k)$. Thus, at each sampling period T_s the current value of the controller's output $u(k)$ is based on the estimates of $\hat{\Theta}(k)$ and set-point signal $r(k)$.

The direct approach ignores the feedback and identifies the LTI system $G_0(q^{-1};\ \Theta)$ using the data set

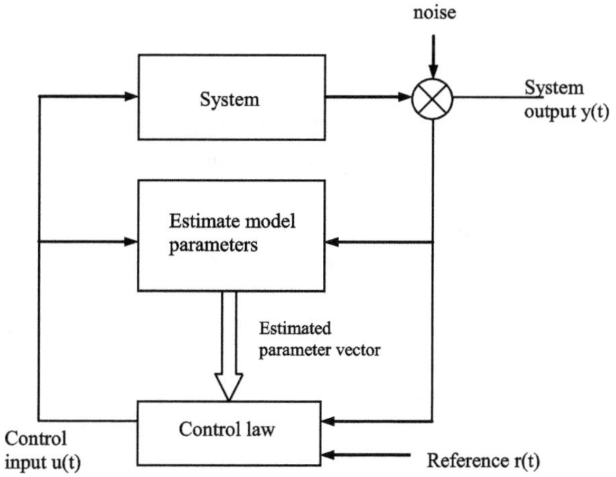

Fig. 6.2 Parameter adaptive self-organizing system with set-point signal $r(k)$

$$Z_s^T = \{u(1), y(1), ..., u(s), y(s)\}, \tag{6.45}$$

containing s pairs of input $u(k)$ and output $y(k)$ samples. By the direct parametric identification method one has to determine the prediction error estimate, using criterion

$$\hat{\Theta}_s = arg \min_{\Theta \in D_M} V_s(\varphi; \theta; \Theta, Z_s). \tag{6.46}$$

Here

$$V_s(\varphi; \theta; \Theta, Z_s) = \frac{1}{s} \sum_{k=1}^{s} e_F^T(k, \varphi, \theta; \Theta) \Lambda^{-1} e_F(k, \varphi; \theta; \Theta), \tag{6.47}$$

with

$$e_F(k, \varphi; \theta; \Theta) = \mathcal{L}(q^{-1}, \theta) \epsilon(k, \varphi; \Theta),$$

$$\epsilon(k, \varphi; \Theta) = H_0^{-1}(q^{-1}, \varphi)(y(k) - G_0(q^{-1}; \Theta) u(k)), \tag{6.48}$$

which are prediction errors; Λ is a symmetric, positive definite weighting matrix and $\mathcal{L}(q^{-1}, \theta)$ is a monic prefilter that can be used to enhance certain frequency regions [5].

The prediction error is calculated by

$$\epsilon(k, \Theta) = y(k) - \hat{y}(k \mid \Theta) = H^{-1}(q^{-1};\ \varphi)[y(k) - G_0(q^{-1};\ \Theta) u(k)], \tag{6.49}$$

where $\hat{y}(k \mid \Theta)$ is the output of the general model of the SISO LTI system $G_0(q^{-1};\ \Theta)$ with noise filter $H_0(q^{-1};\ \varphi)$, respectively.

Then, the one-step-ahead predictor is

$$\hat{y}(k \mid \Theta) = H_0^{-1}(q^{-1};\ \varphi) G_0(q^{-1};\ \Theta) u(k) + [\mathbf{I} - H_0^{-1}(q^{-1};\ \varphi)] y(k), \tag{6.50}$$

where $\mathbf{I}$ is a unitary matrix.

The unknown parameters Θ can be estimated by the well-known PLR algorithm [11]. The filter produces the estimate of the parameter vector $\Theta(k)$ of the form

$$\hat{\Theta}^T(k) = [\hat{\mathbf{a}}^T(k), \hat{\mathbf{b}}^T(k)] = [\hat{a}_1(k), \ldots, \hat{a}_m(k), \hat{b}_1(k), \ldots, \hat{b}_m(k)], \tag{6.51}$$

minimizing the sum of squares of the difference between the generalized model output and plant over some period of time. It requires observations $u(1), u(2), ..., u(N)$ of the input signal $u(k)\ \forall\ k = \overline{1, N}$ and $y(1), y(2), ..., y(N)$ of the output $y(k)\ \forall\ k = \overline{1, N}$ to be processed. The output $y(k)$ turns out into the signal $y_w(k)$, appearing in Eqs. (6.45), (6.48), (6.49) and (6.50), when the linear block $G_0(q^{-1};\ \Theta(k))$ followed a static nonlinearity $f(\cdot, \eta)$.

The direct approach is worked out for a closed-loop LTI system, but not for a nonlinear one. As a consequence, parameter estimate values (6.51) will be far away from their true values $\Theta^T = (\mathbf{a}^T, \mathbf{b}^T) = (a_1, \ldots, a_m, b_0, b_1, \ldots, b_m)$. Hence, there appears a chance to use the direct approach to parameter estimation, if respective inverse filter is inserted into flowchart (Fig. 6.1) in order to diminish the activity of a static nonlinearity. However, an inverse filter that appears in a noisy frame would be the source of errors, intensity of which depends on the activity of additive noises $w(k)$ and $v(k)$.

6.6 Simulation of Nonlinear ARMA System Control Action

6.6.1 Hyperbolic Tangent as a Static Nonlinearity

In an unnoisy frame the linear block of the block-oriented nonlinear system is represented by the second order ARMA model of the form

$$[1 + a_1(k)q^{-1} + a_2(k)q^{-2}]x(k) = [b_1(k)q^{-1} + b_2(k)q^{-2}]u(k). \qquad (6.52)$$

with the initially stationary coefficients: $b_1 = 1, b_2 = 0.5, a_0 = 1, a_1 = -1.5, a_2 = 0.7\ \forall\ k = \overline{1, 400}$. Thus, an unmeasurable intermediate signal is calculated by

$$x(k) = q^{-1}(1 + 0.5q^{-1})(1 - 1.5q^{-1} + 0.7q^{-2})^{-1}u(k). \qquad (6.53)$$

True values of parameters of the linear block (6.52) are:

$$b_1 = 1,\ b_2 = 0.5,\ a_1 = -1.5,\ a_2 = 0.7. \qquad (6.54)$$

In such a case, we will use performance function (6.29) that could be rewritten in an extended form (6.30). By minimizing performance function (6.29) we get two MV controllers. The MV controller for the LTV system without any nonlinear distortion is

$$u(k) = \frac{a_1x(k) + a_2x(k-1) - b_2u(k-2) + r(k)}{b_1}, \qquad (6.55)$$

if the true values of b_1, b_2, a_1, a_2 are known, and

$$u(k) = \frac{\hat{a}_1x(k) + \hat{a}_2x(k-1) - \hat{b}_2u(k-2) + r(k)}{\hat{b}_1}, \qquad (6.56)$$

on the contrary. Here $\hat{b}_1 \equiv \hat{b}_1(k), \hat{b}_2 \equiv \hat{b}_2(k), \hat{a}_1(k) \equiv \hat{a}_1(k), \hat{a}_2 \equiv \hat{a}_2(k)$ are current values of the linear system (block) parameter estimates that are calculated at the k-th recursive iteration; $\hat{b}_1(k) \neq 0$.

Flowcharts of Hammerstein and Wiener systems are shown in Figs. 6.3, 6.4, respectively. In the beginning the simulation was performed without the nonlinearity. As reference sequences (set-point signals), $r(k)$ were used the pulses signal and

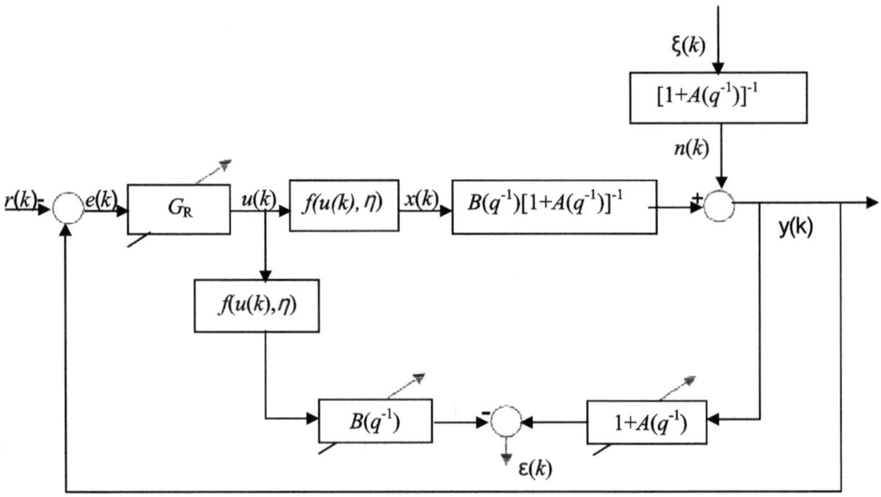

Fig. 6.3 The parametrical identification of the Hammerstein system in a common MV STC flowchart. Signal current values: $r(k)$ (set-point), $e(k)$ (control error), $u(k)$ (control input), $x(k)$ (intermediate), $n(k)$ (disturbance), $y(k) \equiv y_h(k)$ (measured control output), $\xi(k) \sim \mathcal{N}(0, \sigma_\xi)$(white noise), ε (prediction error). G_R is a MV ST controller

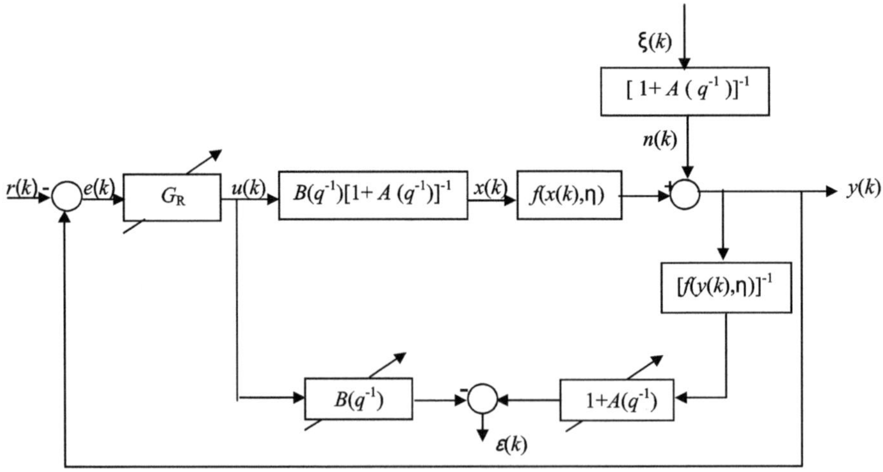

Fig. 6.4 The parametrical identification of the Wiener system in a common MV STC flowchart. Signal current values: $r(k)$ (set-point), $e(k)$ (control error), $u(k)$ (control input), $x(k)$ (intermediate), $n(k)$ (disturbance), $y(k) \equiv y_w(k)$ (measured control output), $\xi(k) \sim \mathcal{N}(0, \sigma_\xi)$ (white noise), ε (prediction error). In flowcharts both signals $x(k)$ do not coincide

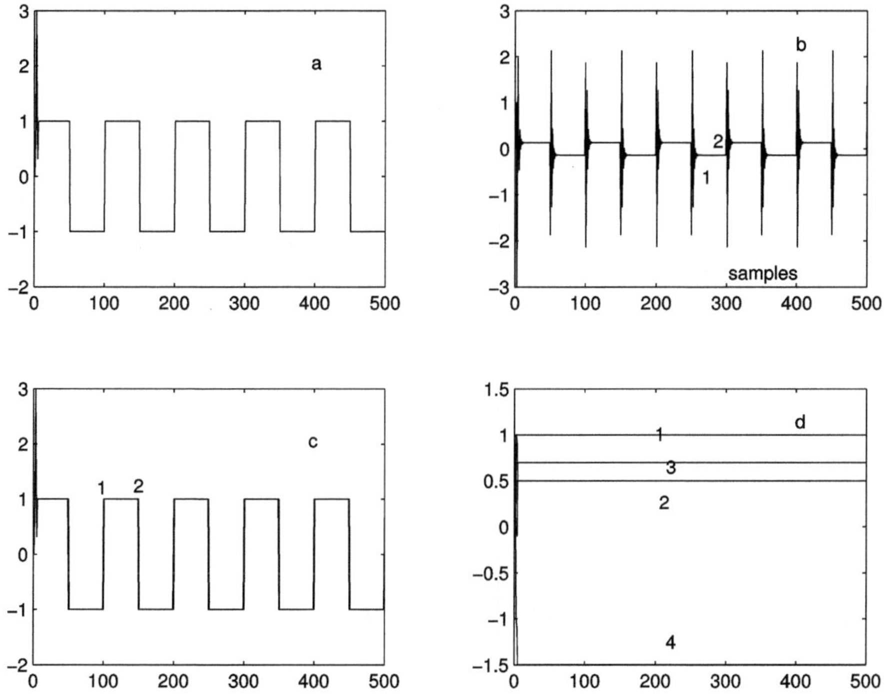

Fig. 6.5 MVST control of the LTI system in dependence of processed samples. Signals: set-point (**a**); control signal (true one (1)), (estimated control signal (2)) (**b**); in (**c**): the same set-point signal from part (**a**) (1), output signal (2). Parameter estimates are coincidental with respective their true values (see **d**)

the stepped signal (Figs. 6.5a and 6.6a). In each recursion, firstly, current value $u(k)$ was calculated by (6.56), secondly, that of $x(k)$ was determined by (6.53), replacing true parameters by their respective estimates. N = 1000 data samples of unnoisy output signals of system have been generated for each input signal, using Eq. (6.53). Afterwards, the self-tuning MV servo control of the initial LTI system was carried out (see Fig. 6.5 and 6.6). The accuracy of servo control was estimated visually, determining how closely the output signal $y(k)$ (really $x(k)$) tracks the given set-point signal $r(k) \forall\, k = \overline{1, 1000}$. Two experiments were performed with distinct signals $r(k)$ (Figs. 6.5a and 6.6a). Corresponding control sequences (6.55) (curve 1), and (6.56) (curve 2) are shown in (Figs. 6.5b and 6.6b), respectively. The reference signal $r(k)$ is given in (Figs. 6.5c and 6.6c) again, as curve 1, where the control output sequence $y_w(k)$ is shown as curve 2, too. The estimates $\hat{b}_1(k)$(curve 1), $\hat{b}_2(k)$(curve 2), $\hat{a}_1(k)$(curve 4), $\hat{a}_2(k)$ (curve 3) of parameters $b_1 = 1, b_2 = 0.5, a_1 = -1.5, a_2 = 0.7$, respectively, are represented in Figs. 6.5d and 6.6d).

The simulation results, given in Figs. 6.5 and 6.6, show us that the MV servo control of LTI system is efficient in an unnoisy frame. The output $y(k)$ is tightly coincidental with the corresponding reference signal $r(k)\ \forall\, k = \overline{1, 1000}$ (Figs. 6.5

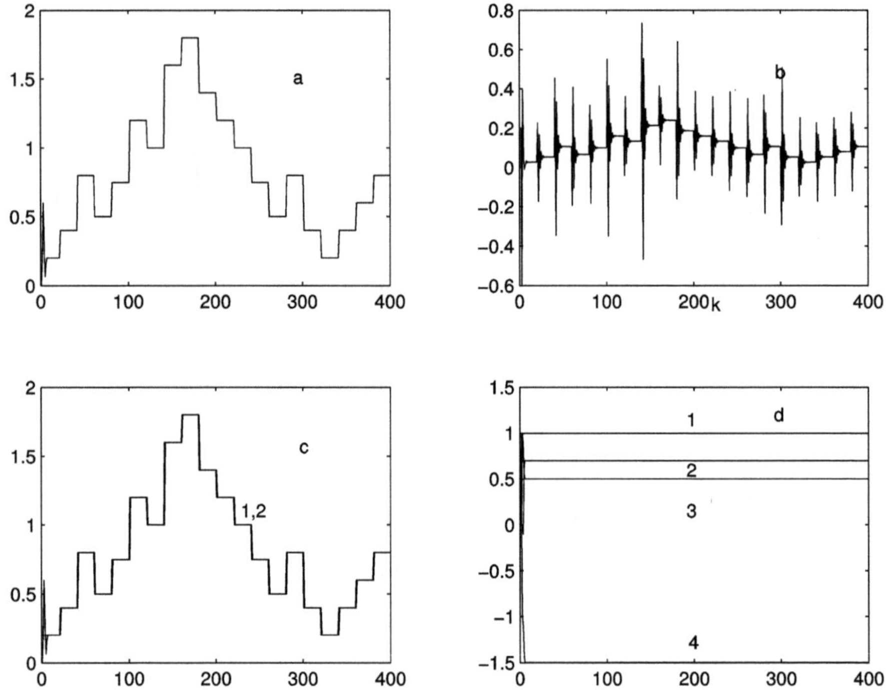

Fig. 6.6 The notation, marking and values are the same as in Fig. 6.5. Only the set point signal (part **a**) is different

and 6.6 parts (a), (c)), as well as coincidental are both control sequences, (1) and (2) (Figs. 6.5 and 6.6 part (b)). One control sequence was obtained by Eq. (6.55), while the other was calculated by (6.56) one. Parameter estimates $\hat{b}_1$, $\hat{b}_2$, $\hat{a}_1$, $\hat{a}_2$ are shown in Figs. 6.5 and 6.6 part (d). In the part (d) curves are: $\hat{b}_1(k)$ (1), $\hat{b}_2(k)$ (2), $\hat{a}_1(k)$ (3), $\hat{a}_2(k)$ (4). Here curve 3 is coincidental with axis X. The parameter estimates $\hat{b}_1$, $\hat{b}_2$, $\hat{a}_1$, $\hat{a}_2$ closely approach true parameter b_1, b_2, a_1, a_2 values, respectively, when the common number of the processed samples is increasing.

Let us suppose now that LTI system (6.52), really, is a nonlinear one. Its linear block follows the static nonparametrical nonlinearity, which is realized by the `Matlab` hyperbolic tangent routine `tanh`. Therefore, the output of the block-oriented nonlinear system is

$$y_w(k) = \mathtt{tanh}\{x(k)\}. \tag{6.57}$$

In the presence of nonlinearity (6.57) the factual form of sequence (6.56), generated by the self-tuning MV controller (Figs. 6.7b and 6.8b), does not correspond to the form of the true control sequence. Curves 1 and 2 in (Figs. 6.7c and 6.8c) are not coincidental. Here, the output control signal $y_w(k)$ (curve 2) does not track the set-up signal $r(k)$ (curve 1) obviously. Decrease of the accuracy, and loss of the efficiency

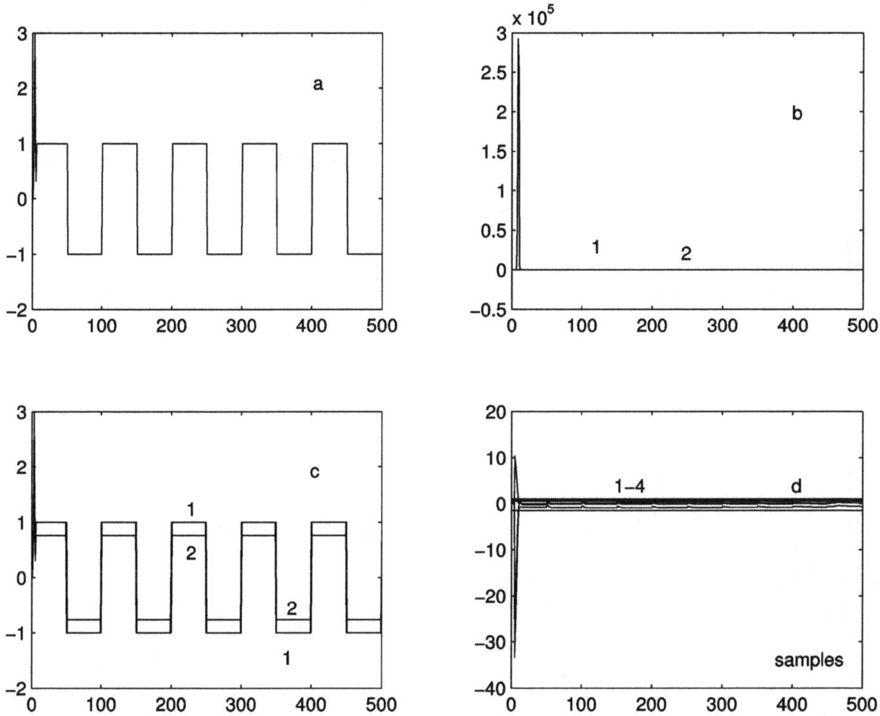

Fig. 6.7 The notation, marking and values are the same as in Fig. 6.5. The non-fixed nonlinearity `tanh` is present

of the MV control are based on strong discrepancy among the true parameter values b_1, b_2, a_1, a_2 and estimated parameters values $\hat{b}_1, \hat{b}_2, \hat{a}_1, \hat{a}_2$ of the linear block (Figs. 6.7d and 6.8d). To increase the accuracy of MV controller is necessary to incorporate in the closed-loop flowchart, as nonlinear filter the inverse hyperbolic tangent, which in Matlab is represented by the routine `atanh` (see Figs. 6.9 and 6.10), respectively. Note that in parts (b) and (d) of Fig. 6.7 the lack of accuracy is not visible due to relatively large discrepancy among some curves. The direct method of the parametrical identification of the closed-loop system will be invalid without `atanh` at the output (see parts (d) of Figs. 6.7 and 6.8) of the flowchart.

It is possible for the Wiener system to compensate distortions, invoked by the static nonlinearity `tanh`, if, at each current iteration, system's output $y_w(k)$ samples were processed by the inverse nonlinearity `atanh`. It could be noted by comparing Figs. 6.9 and 6.10 with Figs. 6.7 and 6.8, respectively. After filtering the control output signal $y_w(k)$, curves 1, 2 in parts (b), and (c) become coincidental (Figs. 6.9 and 6.10). In such a case, the linear part coefficient estimates $\hat{b}_1(k)$(curve 1), $\hat{b}_2(k)$(curve 2), $\hat{a}_1(k)$(curve 4), $\hat{a}_2(k)$ (curve 3) approach the true coefficient values $b_1 = 1, b_2 = 0.5, a_1 = -1.5, a_2 = 0.7$ (parts d of the same figures) speedy and without any oscillations.

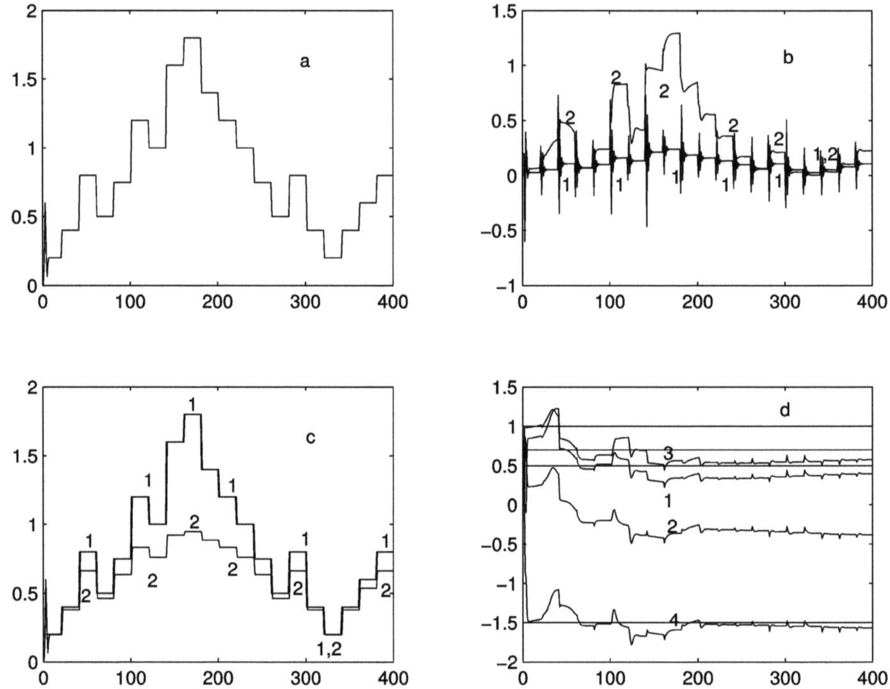

Fig. 6.8 The notation, marking and values are the same as in Fig. 6.6. The non-fixed nonlinearity `tanh` is present

The results given for the Wiener system in Figs. 6.7, 6.8, 6.9 and 6.10 could be compared with the respective results for the Hammerstein system

$$[1 + a_1(k)q^{-1} + a_2(k)q^{-2}]x(k) = [b_0(k)q^{-1} + b_1(k)q^{-2}]\mathtt{tanh}\{u(k)\}, \quad (6.58)$$

represented in Figs. 6.11, 6.12, 6.13 and 6.14. Note that Figs. 6.7 and 6.8 correspond to Figs. 6.11 and 6.13, while Figs. 6.9 and 6.10 correspond to Figs. 6.12 and 6.14, respectively. In (6.58), filtering of the control signal $u(k)$ by the direct nonlinearity `tanh` is used at each current recursive iteration. Thus, to avoid nonlinear distortions for Hammerstein system, it is necessary to calculate values $\mathtt{tanh}\{u(k)\}$, in order to substitute them into Eqs. (6.56) and (6.58).

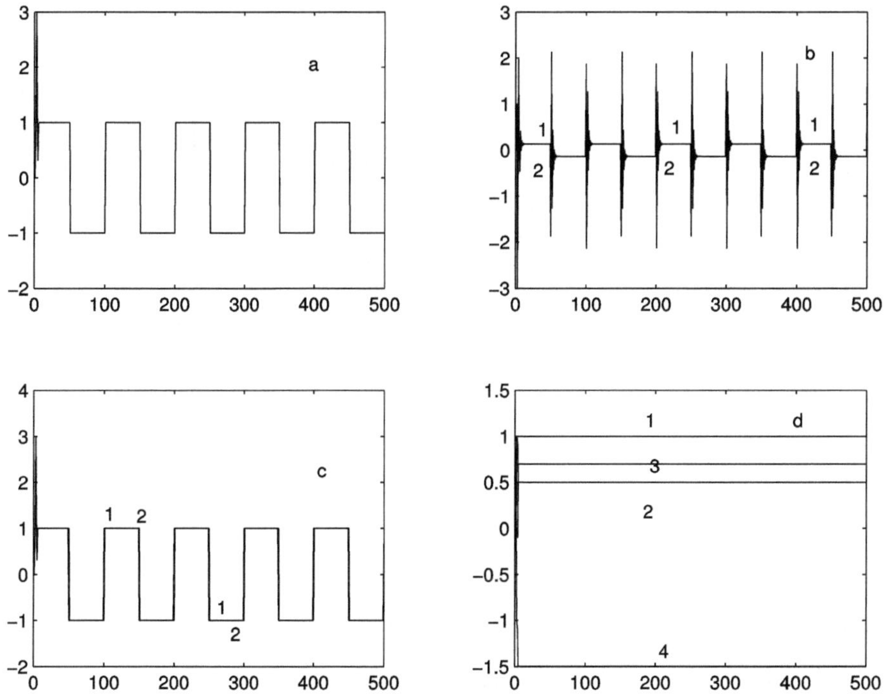

Fig. 6.9 The notation, marking and values are the same as in Fig. 6.7. The inverse nonlinearity `atanh` is used to diminish a parasitical influence of `tanh`

6.6.2 PWL Function as a Static Nonlinearity

In an unnoisy frame the ARMAX system is described by a difference equation (6.52). An unmeasurable intermediate signal is calculated using (6.53). The true values of parameters of the linear block are given in (6.54). The output of the Wiener system is

$$y_w(k) = \begin{cases} -c_0 + c_1 x(k) & \text{if } x(k) \le -d \\ x(k) & \text{if } -d < x(k) \le d, \\ c_0' + c_1' x(k) & \text{if } x(k) > d \end{cases} \tag{6.59}$$

where $c_0 = -0.9$, $c_1 = 0.1$; $c_0' = 0.9$, $c_1' = 0.1$. The threshold $d = 1$.

As reference signals $r(k)$ the periodical sequence of rectangular pulses, and that of staircase form are chosen as sure as in the previous chapter. Note that the three segment function $x(k) = (f(x(k), \eta) - c_0)/c_1$, if $x(k) \le -a$, and $x(k) = f(x(k), \eta)$, if $-a < x(k) \le a$, and $x(k) = (f(x(k), \eta) - c_0')/c_1'$, if $x(k) > a$ build up the PWL function. The nonlinear Hammerstein system is described by a difference equation

$$(1 + a_1(k)q^{-1} + a_2(k)q^{-2})y_H(k) = (b_0(k)q^{-1} + b_1(k)q^{-2})\varphi(u(k), \eta), \tag{6.60}$$

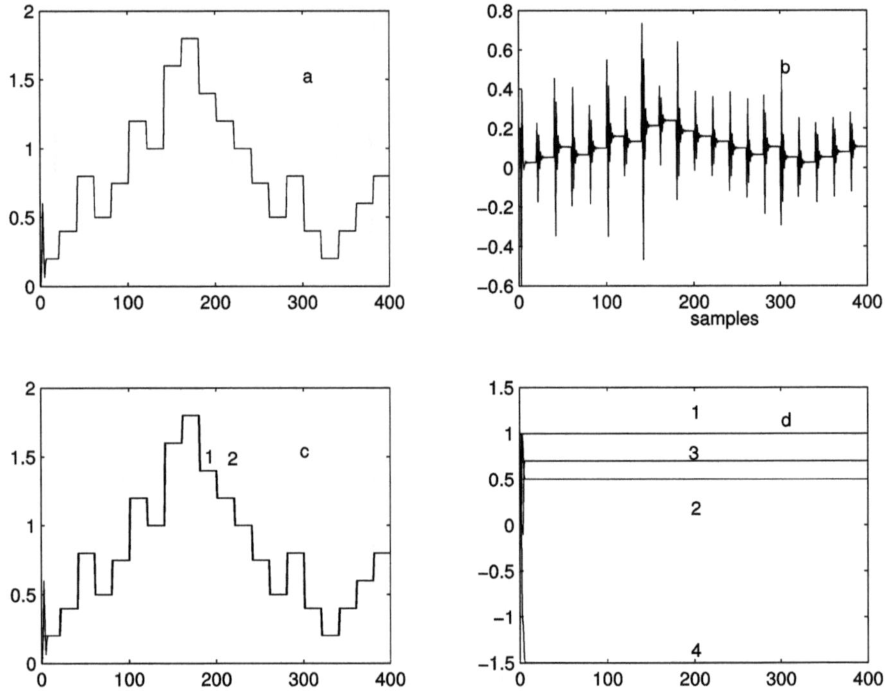

Fig. 6.10 The notation, marking and values are the same as in Fig. 6.8. The inverse nonlinearity `atanh` is used to diminish the influence of `tanh`

having in mind that the linear block is preceded by a static PWL nonlinearity (6.59). The initial values of the vector $\hat{\Theta}^T(0)$ and the matrix $\Gamma(0)$ are:

$$\hat{\Theta}^T(0) = (0, 0, 0, 0), \tag{6.61}$$

$$\Gamma(0) = \begin{bmatrix} 10^5 & 0 & 0 & 0 \\ 0 & 10^5 & 0 & 0 \\ 0 & 0 & 10^5 & 0 \\ 0 & 0 & 0 & 10^5 \end{bmatrix}. \tag{6.62}$$

The parameters c_0, c_1 and d_0, d_1 were calculated, when $y(k) \leq -1$ and when $y(k) > 1$, in dependence on the switching function:

$$h(\alpha) = \begin{cases} 0 \text{ if } \alpha \geq 0, \\ 1 \text{ if } \alpha < 0. \end{cases} \tag{6.63}$$

It switches between two sets of values, i.e.,$(-\infty, \alpha)$ and (α, ∞), where $h(\alpha) = h(d - x(k))$, in the first case, and $h(\alpha) = h(x(k) - d)$ in the second one.

Calculations are performed by processing the current pair of samples $u(k)$ and $y(k)$.

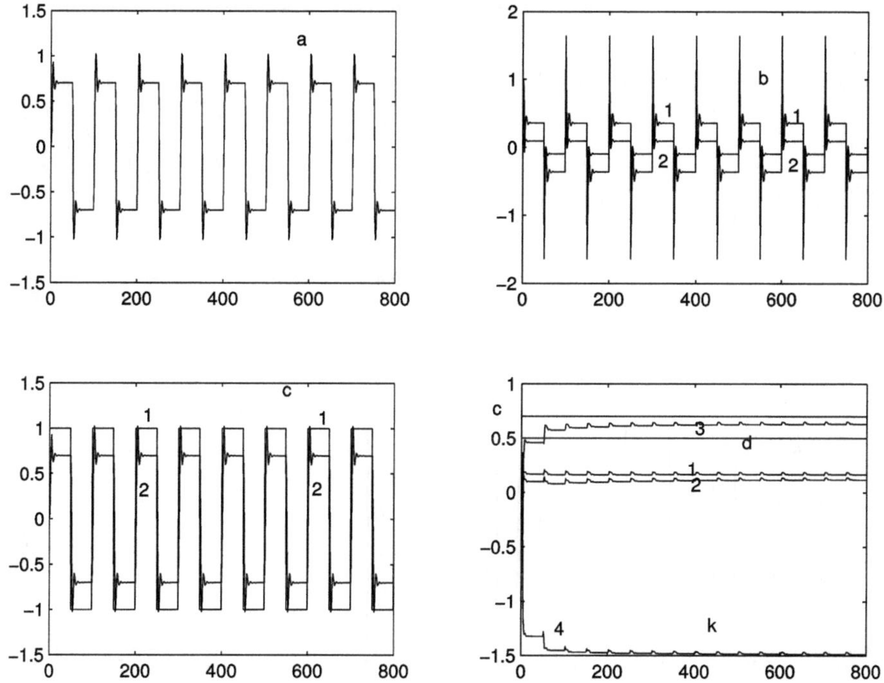

Fig. 6.11 MV self-tuning control of the Hammerstein system (6.58) in dependence of processed samples. The nonlinearity 'tanh' is fixed. The notation, marking and values are the same as in Fig. 6.7

The GMV ST controller for the system of II order is

$$u(k) = \frac{p(a_1 y(k) + a_2 y(k-1) - b_2 u(k-1)) + g r(k)}{p b_1 + q}. \tag{6.64}$$

Here $u(k)$ is the current control signal value; p, q, and g are weighting coefficients. If p=g=1 and q=0, the controller is a MV controller.

To implement the ST MV controller correctly, it is necessary, firstly, in the learning phase, to obtain the value of the threshold d by reordering the observations and analyzing the corresponding plots, i.e., Fig. (1.4). Secondly, in each current iteration to calculate the estimate of parameters b_1, b_2, a_1, a_2 by the PLR algorithm.

It is known [8] that for the open-loop system parametric identification is efficient only if the transfer function is

$$H_0(q^{-1}; \theta) = \frac{1}{1 + A(q^{-1}; \mathbf{a})} = \frac{1}{1 + a_1 q^{-1} + \cdots + a_{n_a} q^{-n_a}}, \theta \equiv \mathbf{a}. \tag{6.65}$$

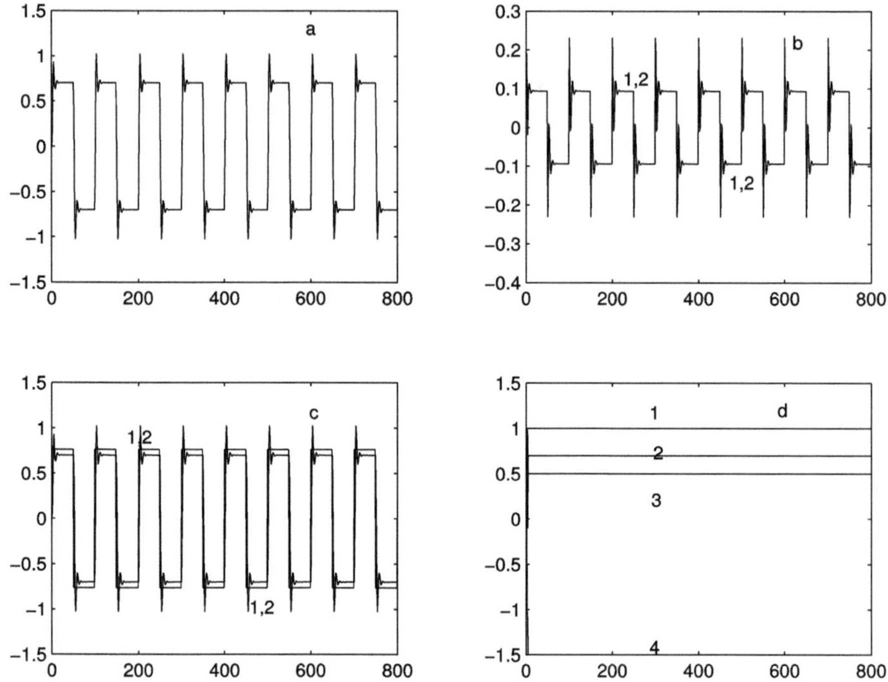

Fig. 6.12 The notation, marking and values are the same as in (see Fig. 6.9). The nonlinearity 'tanh' is fixed (see Fig. 6.9). Satisfactory results are obtained, by substituting the values $\texttt{tanh}\{u(k)\}$ instead of $u(k)$ (see Fig. 6.11) into vector **Z** Eq. (6.45)

In a case of a closed-loop system the respective identifiability conditions should be satisfied before the parametric identification [9].

Five distinct experiments, based on flowcharts Figs. 6.3, and 6.4, have been carried out in an unnoisy frame.

Experiment 1 was performed without the PWL nonlinearity [10]. N = 1000 data samples of unnoisy input-output signals have been generated. The ST MV control of the initial LTI system was carried out (see Fig. 6.15). The reference and controlled output signals $r(k)$, $y_w(k)$ are shown in Fig. 6.15 part **a**, respectively. The true control signal calculated by (6.55), and its estimate, obtained by (6.56), are given in the part (b). Note that the reference signal $r(k)$ is given twice. In part (**a**) the signal $r(k)$ is represented together with the control output $y_w(k)$, while in part (c) it is given separately. Parameter estimates $\hat{b}_1, \hat{b}_2, \hat{a}_1, \hat{a}_2$ are shown in part (d). The curves shown in part (d): $\hat{b}_1(k)$ (1), $\hat{b}_2(k)$ (2), $\hat{a}_1(k)$ (3), $\hat{a}_2(k)$ (4). Here curve 3 is coincidental with axis X. The same disposition of signals and parameters as well as their estimates are unbroken in Figs. 6.16 and 6.17, too.

Simulation results, given in Fig. 6.15, show us that the MV control of the LTI system in an unnoisy environment is efficient. The output $y_w(k)$ coincides with the reference signal $r(k)$ $\forall\, k = \overline{1, 1000}$ (Fig. 6.15, part (a), as well as both control

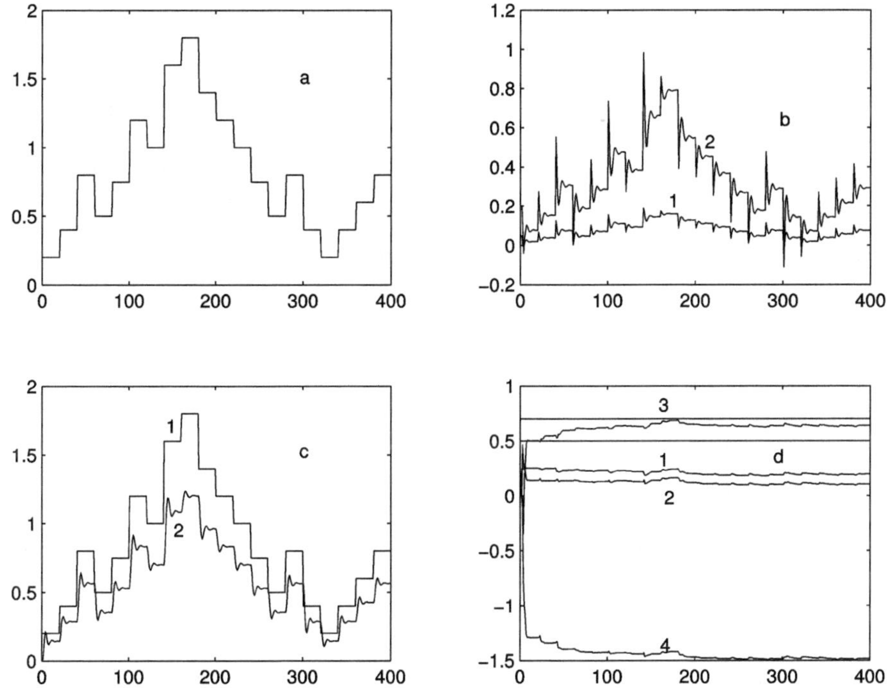

Fig. 6.13 The notation, marking and values are the same as in (see Fig. 6.8)

sequences, true sequence (1) obtained in Eq. (6.55), and estimated one in (2), are calculated in (6.56) (part (b)). Parameter estimates $\hat{b}_1, \hat{b}_2, \hat{a}_1, \hat{a}_2$ approach the true parameter values b_1, b_2, a_1, a_2, respectively, when the common number of the processed samples (Fig. 6.15 part (d)) has been increased.

Experiment 2 was provided for the PWL nonlinearity (6.59) in the flowchart of the closed-loop LTI system. Nevertheless, the presence of nonlinearity (6.59) was not registered. The output of the Wiener system $y_w(k)\ \forall\ k = \overline{1, 1000}$ was generated by the PWL function with positive slopes (6.59). A control of the nonlinear block-oriented Wiener system was carried out, assuming that the controlled system is the LTI one. That is why at each current iteration k, only the estimates of coefficients of linear block (6.52) were calculated (Fig. 6.16 part (d)). They are used to find the current value of control sequence (6.56). The simulation results in (Fig. 6.16) show us that there exists a great discrepancy between ideal and real signals (parts: (a), (b)). The estimates of coefficients b_1, and b_2 are far away from their true values, even for the increased iteration numbers. Therefore, we can state that the MV control of LTI system (6.52) will be non-efficient even in the unnoisy frame if the nonlinearity, available in the system control flowchart, is not registered.

It was assumed in **Experiment 3** that parameter values of nonlinearity (6.59) are known exactly (Fig. 6.17). Therefore, the current value of output of the Wiener system

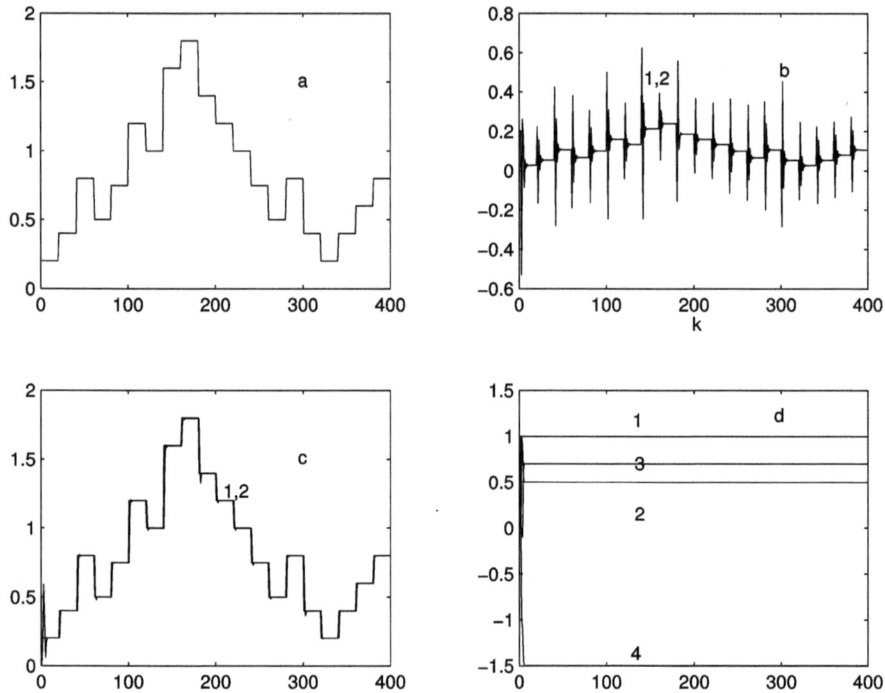

Fig. 6.14 The notation, marking and values are the same as in (see Fig. 6.10). Satisfactory results are obtained, substituting the values $\mathtt{tanh}\{u(k)\}$ instead of $u(k)$ (see Fig. 6.13) in vector $\mathbf{Z}$ Eq. (6.45)

$y(k)\ \forall\ k = \overline{1, 1000}$ in each iteration k was passed through the invertible nonlinear filter of the form

$$f(x(k), \eta)^{-1} = \begin{cases} (y(k) + 0.9)/0.1 & \text{if } y(k) \leq -1, \\ y(k) & \text{if } -1 < y(k) \leq 1, \\ (y(k) - 0.9)/0.1 & \text{if } 1 < y(k) \end{cases} \tag{6.66}$$

$\forall\ k = \overline{1, 1000}$, eliminating the parasitic influence of the static PWL nonlinearity (6.59). There were only estimates of parameters of the linear block calculated (Fig. 6.17d). In such a case, the simulated and estimated signals (parts (a), (b)) are coincidental. The estimates of linear block (6.52) coefficient values are very close to their true values, respectively (part (d)).

Experiment 4. It was supposed here that the parameters of PWL nonlinearity (6.59), as well as that of the linear block (6.52) are unknown. Therefore, it is necessary to calculate their estimates. The current value of the output of the Wiener system $y_w(k)\ \forall\ k = \overline{1, 1000}$ in each iteration was passed through the invertible nonlinear filter of the form

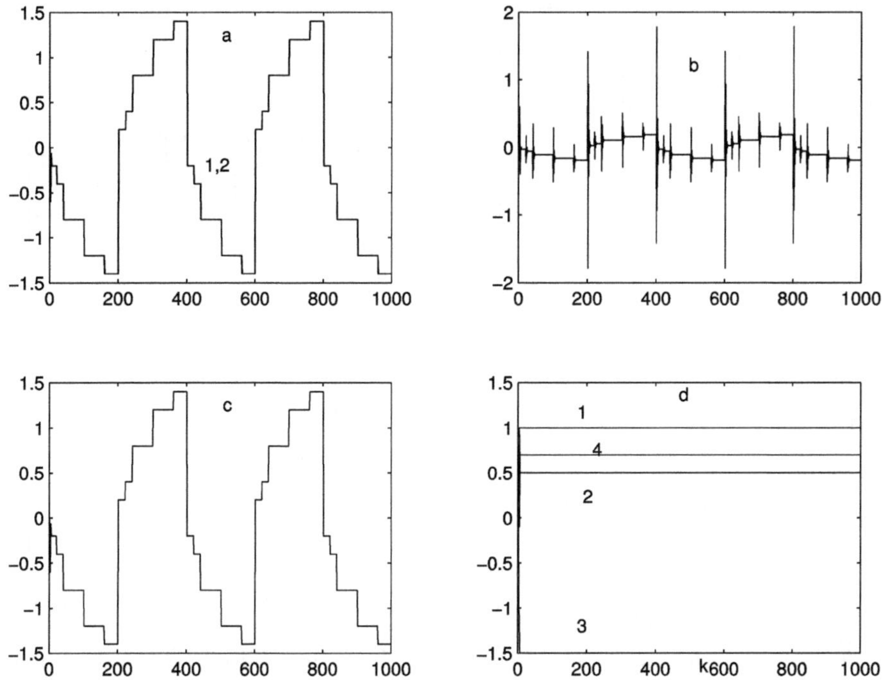

Fig. 6.15 Signals (parts: (**a**), (**b**), (**c**)) and parameter estimates of the self-tuning LTI system (part (**d**)) depending on number of iterations k

$$f(x(k), \hat{\eta})^{-1} = \begin{cases} (y(k) + \hat{c}_0(k))/\hat{c}_1(k) & \text{if } y(k) \leq -1, \\ y(k) & \text{if } -1 < y(k) \leq 1, \\ (y(k) - \hat{c'}_0(k))/\hat{c'}_1(k) & \text{if } 1 < y(k), \end{cases} \tag{6.67}$$

$\forall\, k = \overline{1, 1000}$, where their estimates were substituted instead of true and unknown parameters.

Here $\hat{\eta} \equiv \hat{\eta}(k)$=$(\hat{c}_0(k), \hat{c}_1(k), \hat{c'}_0(k), \hat{c'}_1(k), d)^T$ is the vector of parameter estimates of the PWL nonlinearity (6.59).

In Fig. 6.18, signals and parameter estimates of the Wiener system, controlled with STC are given. In part (a) curves 1 and 2, that correspond to $r(k)$ and $y(k)$ are shown, respectively. The true control signal calculated in (6.55), and its estimate obtained in (6.56) are given in part (b). The parameters of nonlinearity (6.66) were simultaneously estimated with the parameters of the linear block. Note that in Fig. 6.17c curves correspond to: 1 ($\hat{c}_0(k)$), 2 ($\hat{c}_1(k)$), 3($\hat{c'}_0(k)$), 4 ($\hat{c'}_1(k)$). Curves in part (d) are following: 1 ($\hat{b}_1(k)$), 2 ($\hat{b}_2(k)$), 3 ($\hat{a}_1(k)$), 4 ($\hat{a}_2(k)$). It is obvious that the accuracy of parameter estimates is not sufficient to adjust the control signal $y_w(k)$ properly.

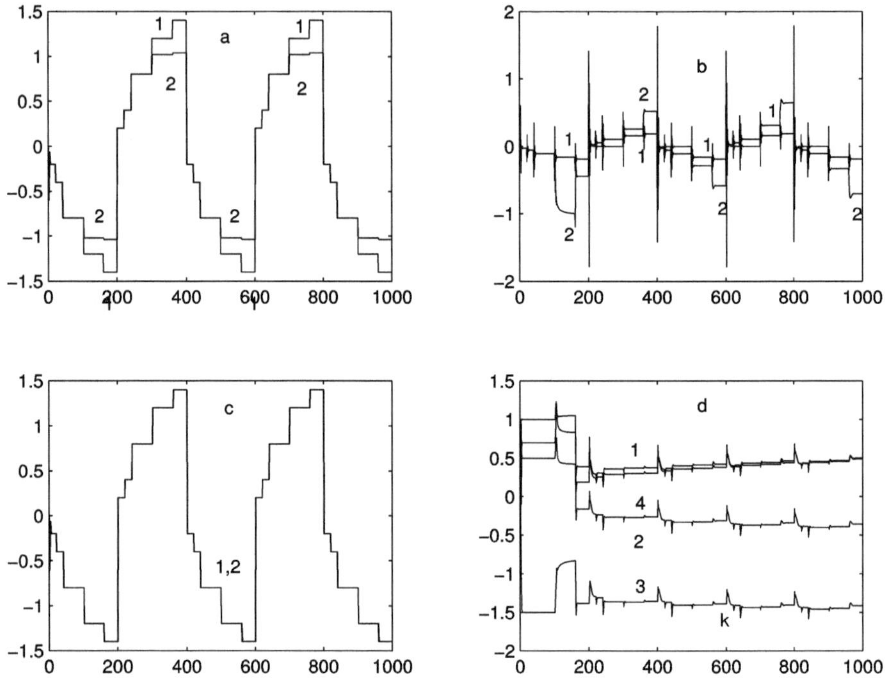

Fig. 6.16 Signals are (parts: (**a**), (**b**), (**c**)), and parameter estimates of the LTI system (part (**d**)). The nonlinearity is present but not registered. Marks, notation, and values are the same as in Fig. 6.15

In **Experiment 5** the MV controller was changed by the GMV one, fixing the influence on the control quality. The simulation results, given in Fig. 6.19, do not allow us to state that the GMV method gives us a better control quality, as compared with the MV method (Fig. 6.18). Moreover, there appeares an enormous spike in Fig. 6.18a. The spike usually invokes the unstability of the MV controller. Fortunately, alone spike did not knock down the stability of the control system significantly. Such or similar spikes are absent in Fig. 6.19a. However, both methods do not guarantee the satisfactory accuracy of parameter estimates of the linear block (6.52) as well as that of PWL nonlinearity (6.59) (parts (c), (d). Their control laws are swayed because of the nonaccuracy of estimate values.

Finally, it is necessary to emphasize that to control LTI system, having the two, three or more segments PWL nonlinearity, efficiently, it is necessary:

- to know the type (Wiener or Hammerstein) of the nonlinear system;
- to know the form of the nonlinearity i.e., segment slopes, break-points, constants;
- to obtain parameter estimates if parameters of the nonlinearity are unknown;
- to calculate the nonlinearity parameter estimates as close as possible to the respective true values;
- in a case of Wiener system to build an inverse filter, in which instead of true but unknown system parameters their current estimates are substituted;

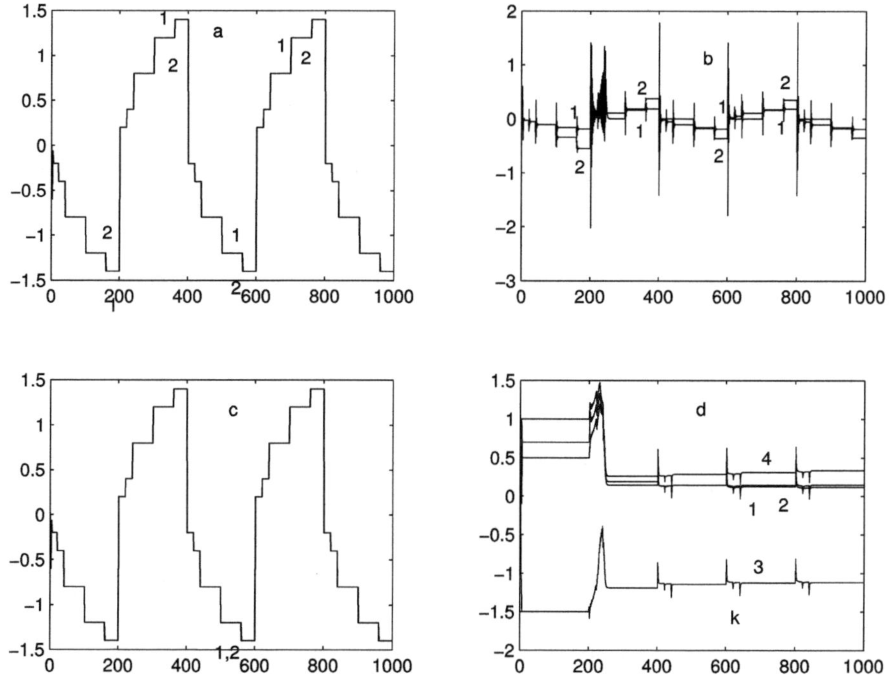

Fig. 6.17 Signals are(parts: (**a**), (**b**), (**c**), and parameter estimates of the Wiener system (part (**d**)). The nonlinearity is present and known in respect of parameters: (c_0), (c_1), $((c'_0)$, (c'_1). The inverse filter (6.66) is used with known parameters. Marks and notation are the same as in Fig. 6.15

- to implement the inverse filter into the control flowchart in order to eliminate the unwanted nonlinear effects during each current k iteration (Fig. 6.4).
- in a case of Hammerstein system to process input signal samples by the direct nonlinearity (Fig. 6.3).

The identifiability conditions given in Isermann [9] should be satisfied as well.

Note that approaches, based on signal sampling with stationary sampling period T_s, were used here, while samples of different regimes of three segment function (6.59) are mixed up ocasionally. Really, in such a situation procedures, efficient to process signals, sampled with time-varying T_s are necessary. Generally, it is not clear how to change the sampling period T_s while MV control. Therefore, the one-shot controller tuning during the training phase, based on the data reordering, can be preferable. In such a case, the blocks, representing adaptive parametric identification and the controller parameters calculation, would be connected at the moment during the adjustment phase. Once the controller has been adjusted, the block of the recursive parametric identification and the block of the controller parameter calculation should be disconnected (Bjobal et al. 2000). Afterwards, the system would be controlled under fixed parameters. This method of control is useful for deterministic processes, corrupted by small intensity noise.

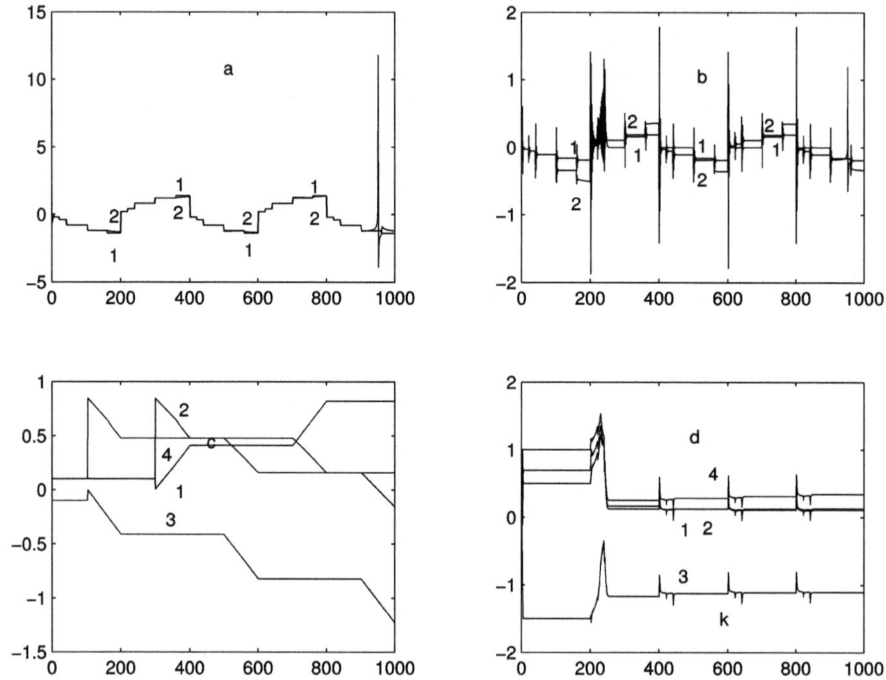

Fig. 6.18 The PWL nonlinearity is unknown in respect of the parameters: (c_0), (c_1), $((c'_0)$, (c'_1). The inverse filter (6.66) is used with the estimates $(\hat{c}_0(k))$, $(\hat{c}_1(k))$, $(\hat{c'}_0(k))$, $(\hat{c'}_1(k))$, which are shown in part **c** as curves 1–4, respectively. The marks and notation are the same as in Fig. 6.15

6.7 Conclusions

The potentiality of nonlinear ARMAX systems for MV control were explored here in order to implement in the control flowchart the MV or GMV controllers (see Fig. 6.1) that are based on the numerical optimization of the performance functions (6.40) and (6.41), respectively. The well-known direct method, which belongs to the class of closed-loop parametric identification methods, is used here to estimate the parameter vector $\Theta(k)$ of the linear block $G_0(q^{-1};\ \Theta(k))$. Both groups of parameters that belong to linear as well as to nonlinear blocks could be calculated by separate recursive routines. The various results of simulation, corroborating the efficiency of nonlinear ARMAX systems for MV control, were obtained.

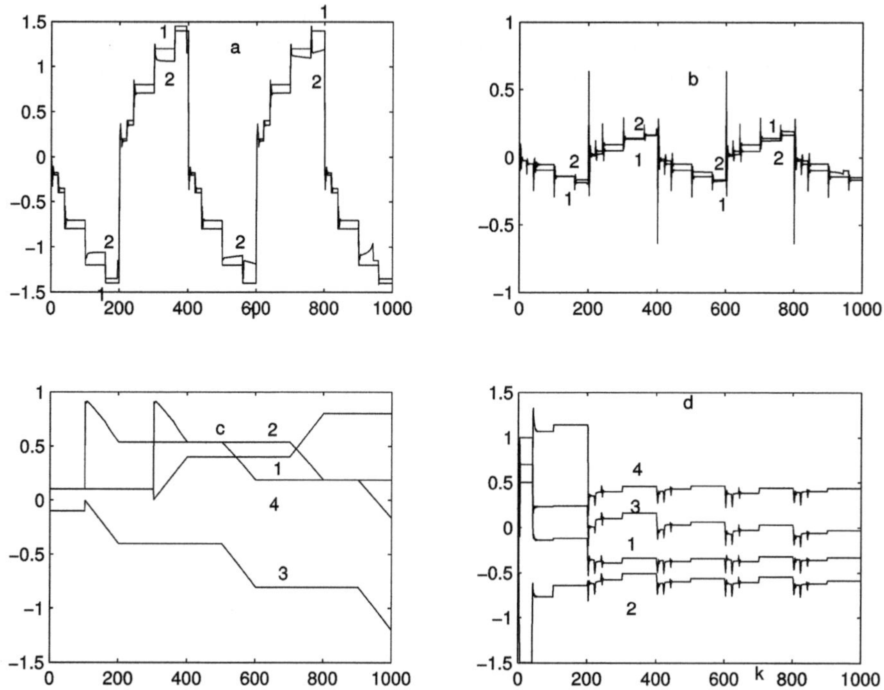

Fig. 6.19 The marks and notation are the same as in Fig. 6.18. For simulation and parametric identification the GMV control law is used instead of MV control law

References

1. Anderson, B.D.O., Moore, J.B.: Linear Optimal Control. Prentice-Hall, Englewood Cliffs, NJ (1971)
2. Anderson, B.D.O., Moore, J.B.: Optimal Filtering. Prentice-Hall, Englewood Cliffs, NJ (1979)
3. Åström, K.J.: Introduction to Stochastic Control Theory. Academic, NY (1970)
4. Bobál, V., Böhm, J., Fessl, J., Machaček, J.: Digital Self-Tuning Controllers. Springer, Berlin (2005)
5. Forssell, U., Ljung, L.: Closed-loop identification revisited. Automatica **35**(1999), 1215–1241 (1999)
6. Goodwin, G.C., Sang Sin, K.: Adaptive Filtering, Prediction and Control. Prentice Hall, Englewood Cliffs, NJ (1984)
7. Evans, G.: Self-tuning and adaptive control coursework (2000). http://www.taumuon.co.uk/cv/SelfTuningAdaptiveControl.doc. Accessed in Sept 2000
8. Isermann, R.: Prozessidentifikation. Springer, Berlin (1974)
9. Isermann, R.: Practical aspects of process identification. Automatica **16**(5), 575–587 (1980)
10. Isermann, R.: Digital Control Systems. Springer, Berlin (1981)
11. Ljung, L.: System Identification—Theory for the User, 2nd edn. Prentice Hall, Upper Saddle River, NJ (1999)
12. Ljung, L., Gunnarsson, S.: Adaptation and tracking in system identification—a survey. Automatica **26**(1), 7–21 (1990)
13. Ljung, L., Söderström, T.: Theory and Practise of Recursive Identification. MIT Press, Cambridge (1983)

14. Mišković, L.M., Djurović Ž, M., Kovačevi, ć.: Application of the minimum state error variance approach to nonlinear system control. Int. J. Syst. Sci. **33**(5), 359–364 (2002)
15. Tham, M.T.: Minimum Variance and Generalized Minimum Variance Control Algorithms. University of Newcastle upon Tyne (1999). http://lorien.ncl.ac.uk/ming/digicont/mbpc/gmv.pdf. Accessed May 1999

Chapter 7
Parametric Identification of Systems with Piecewise-Linear Nonlinearities

It is shown here that by a simple data reordering and by a following data partition the problem of parametrical identification of Wiener and Hammerstein systems with PWL nonlinearity could be reduced to a multifold linear parametric estimation problem. Afterwards, we will represent the former regression function in 'pieces', using ordinary FIR filters and the well-known linear convolution. Consistency of parameter estimates were investigated here, too. Simulation results are given for the Wiener system with invertible as well as noninvertible PWL nonlinearities.

7.1 Statement of the Problem

The linear part of the block-oriented Wiener system is dynamic, time invariant, causal, and stable. It can be represented by a dynamic LTI system with a transfer function $G(q^{-1}, \mathbf{\Theta})$ as a rational invertible function of the form

$$G(q^{-1}, \mathbf{\Theta}) = \frac{b_1 q^{-1} + \cdots + b_m q^{-m}}{1 + a_1 q^{-1} + \cdots + a_m q^{-m}} = \frac{B(q^{-1}, \mathbf{b})}{1 + A(q^{-1}, \mathbf{a})} \tag{7.1}$$

with polynomials

$$A(q^{-1}, \mathbf{a}) = \sum_{i=1}^{n_a} a_i q^{-i}, \quad B(q^{-1}, \mathbf{b}) = \sum_{i=1}^{n_b} b_i q^{-i}, \tag{7.2}$$

having a finite number of parameters

$$\mathbf{\Theta}^T = (b_1, \ldots, b_n, a_1, \ldots, a_n),$$

$$\mathbf{b}^T = (b_1, \ldots, b_n), \quad \mathbf{a}^T = (a_1, \ldots, a_n) \tag{7.3}$$

R. Pupeikis and K. Kazlauskas, *Advanced Digital Signal Processing Methods for Filtering, Identification, and Nonlinear Systems Control*, Studies in Systems, Decision and Control 638, https://doi.org/10.1007/978-3-032-06740-1_7

that are determined from the set $\boldsymbol{\Omega}$ of permissible parameter values $\boldsymbol{\Theta}$. Here q^{-1} is a backward time-shift operator; the set $\boldsymbol{\Omega}$ is restricted by conditions on the stability of the respective difference equation, $n_a = n_b = m$.

The output signal

$$y(k) = f(x(k), \eta) + e(k) \tag{7.4}$$

is generated by the nonlinear part $f(\cdot, \eta)$ of the Wiener system, as a response to the unmeasurable intermediate signal

$$x(k) = \frac{B(q^{-1}, \mathbf{b})}{1 + A(q^{-1}, \mathbf{a})} u(k) + v(k). \tag{7.5}$$

The nonlinear part $f(\cdot, \eta)$ of the Wiener system (Fig. 1.3) is PWL function as follows [4, 5, 10, 15]

$$f(x(k), \eta) = \begin{cases} c_0 - c_1 x(k) & \text{if } x(k) \le -a \\ x(k) & \text{if } -a < x(k) \le a, \\ d_0 - d_1 x(k) & \text{if } x(k) > a \end{cases} \tag{7.6}$$

(Figures 7.1 and 7.2) or the PWL function of the form

$$f(x(k), \eta) = \begin{cases} x(k) - d & \text{if } x(k) \ge d, \\ 0 & \text{if } -d < x(k) < d, \\ x(k) + d & \text{if } x(k) \le d. \end{cases} \tag{7.7}$$

(Figure 7.3) that both could be partitioned into three PWL functions.

For the nonlinearity (7.6) these functions are: $f\{x(k; \boldsymbol{\Theta}), \mathbf{c}, a\} = c_0 - c_1 x(k)$, $f\{x(k; \boldsymbol{\Theta}), a\} = x(k)$, and $f\{x(k; \boldsymbol{\Theta}), \mathbf{d}, a\} = d_0 - d_1 x(k)$. In (7.6): function $f\{x(k; \boldsymbol{\Theta}), \mathbf{c}, a\}$ has only negative values, when $x(k) \le -a$, $f\{x(k; \boldsymbol{\Theta}), a\}$ has arbitrary positive, as well as negative values, when $-a < x(k) \le a$, and $f\{x(k; \boldsymbol{\Theta}), \mathbf{d}, a\}$ has only positive values, when $x(k) > a$. Here $x(k; \boldsymbol{\Theta}) \equiv x(k)$, $\mathbf{c}^T = (c_0, c_1)$, $c_0 = -a(1 + c_1), 0 < c_1 << a, \mathbf{d}^T = (d_0, d_1), d_0 = a(1 + d_1), 0 < d_1 << a$, when $a >$

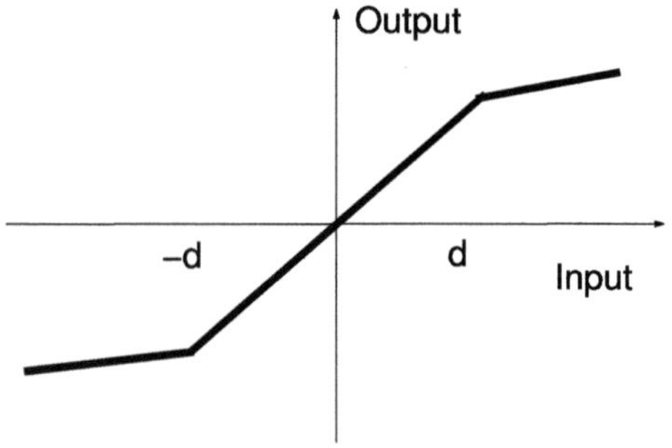

Fig. 7.1 Three segments PWL nonlinearity $f(\cdot, \eta)$ with positive slopes. Here a is a threshold

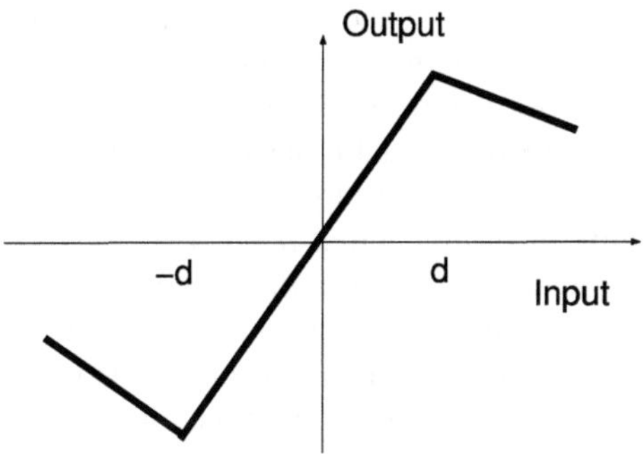

Fig. 7.2 Non-invertible three segments PWL characteristic $f(\cdot, \eta)$

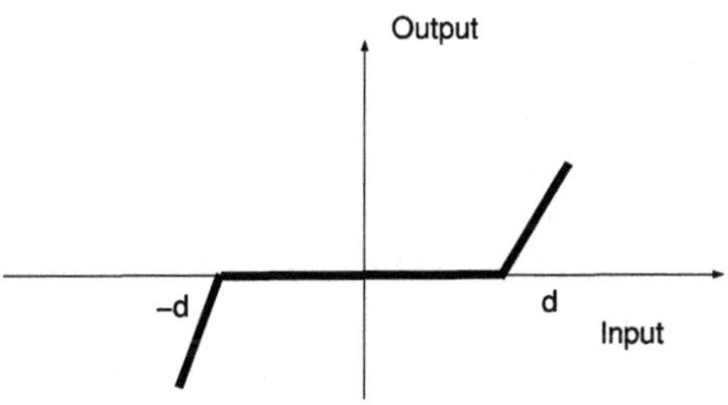

Fig. 7.3 Deadzone nonlinearity $f(\cdot, \eta)$

0. It is simple, having the nonlinearity (Eqs. (7.6) and Fig. 7.1), to obtain the non-invertible PWL nonlinear function (see Fig. 7.2). Corresponding functions can be written for the deadzone (7.7), Fig. 7.3 as well.

The process noise $v(k) \equiv \xi(k)$ and the measurement noise $e(k) \equiv \zeta(k)$ are added to an intermediate signal $x(k)$ and the output $y(k)$, respectively, $\xi(k)$, $\zeta(k)$ are mutually noncorrelated sequences of independent Gaussian variables with $\mathcal{E}\{\xi(k)\} = 0$, $\mathcal{E}\{\zeta(k)\} = 0$, $\mathcal{E}\{\xi(k)\xi(k+\tau)\} = \sigma_\xi^2\delta(\tau)$, $\mathcal{E}\{\zeta(k)\zeta(k+\tau)\} = \sigma_\zeta^2\delta(\tau)$; $\mathcal{E}\{\cdot\}$ is a mathematical expectation, $\sigma_\zeta^2, \sigma_\xi^2$ are variances of ζ and ξ, respectively, $\delta(\tau)$ is the Kronecker delta function.

The objective is to find the way to estimate parameters (7.3) of the linear part (7.1), as well as parameters $\eta = (c_0, c_1, d_0, d_1)^T$ and the threshold a of the available nonlinearity (7.6) by processing N pairs of observations $u(k)$ and $y(k)$ of the block-oriented Wiener system, acting in an open-loop flowchart.

7.2 Data Reordering in a Case of Piecewise-Linear Nonlinearity

The method, based on the signal samples reordering, is proposed here for the determining of unknown parameters. Thus, let us rearrange the open-loop Wiener system (Fig. 1.3) data $y(k)$ $\forall\, k \in \overline{1, \mathrm{N}}$ in an ascending order of their values, assuming that noises $w(k) \equiv \xi(k)$, $v(k) \equiv \zeta(k)$ in Fig. 1.3 are absent, parameters $\eta =$

$(c_0, c_1, d_0, d_1)^T$, and the threshold a of nonlinearity (7.6) are known [11]. The observations of the reordered output $\tilde{y}(k)\ \ \forall\, k \in \overline{1, \mathrm{N}}$ of the Wiener system should be partitioned into three data sets: left-hand side data set (N_1 samples) with values lower than or equal to negative a, middle data set (N_3 samples) with values higher than negative a but lower or equal to a, and right-hand side data set (N_2 samples) with values higher than a, where $\mathrm{N} = \mathrm{N}_1 + \mathrm{N}_3 + \mathrm{N}_2$ is the samples size of the initial sequence $y(k)$. Hence, the observations with the highest and positive values will be concentrated on the right-hand side set, while the observations with the lowest and negative values on the left-hand side one. It could be noted that on boundaries the small portions of observations of the middle data set of $\tilde{y}(k)$ are mixed together with some small portions of observations of the left-hand side and right-hand data sets, respectively, due to positive slopes of the nonlinearity (7.6). In general case (an noisy environment, unknown parameters and the threshold, too) it is imperative for the efficient parametric identification of the block-oriented Wiener system that such ambiguities are resolved. We can find the threshold and respective breakpoints of nonlinearity, calculating the simple dependence of output $y(k)$ on the input $x(k)$ (Eq. (7.6) and Fig. 7.1). Then, the initial PWL regression function, representing abovementioned dependence, could be subdivided into the finite number of pieces, each of which having individual description. However, in a noisy frame the boundaries of different pieces could be not visible clearly. One can avoid this problem assuming here that slightly more than 80% unmixed observations of large enough whole data set N are concentrated on the middle-set $\hat{\mathrm{N}}_3$ and approximately by 10% of the same data set on any side set $\hat{\mathrm{N}}_1$ or $\hat{\mathrm{N}}_2$, where $\hat{\mathrm{N}}_1$, $\hat{\mathrm{N}}_2$ and $\hat{\mathrm{N}}_3$ are estimated values of $\mathrm{N}_1, \mathrm{N}_2$ and N_3, respectively, that are not necessary coincidental with true ones. On the other hand, that observation values of the middle data set of $\tilde{y}(k)$ are coincidental with the respective observation values of the intermediate signal $x(k)$ in the absence of the measurement noise $\upsilon(k)$. Therefore, one could get unmixed observations of $\tilde{y}(k)$ simply by choosing the upper interval bound lower than the 80% and the lower interval bounds higher than the 10% of the sampled reordered observations of $\tilde{y}(k)$.

Next, let us reconstruct an unmeasurable intermediate signal $x(k)$, using the middle data set $\tilde{y}(k)\ \forall\, k \in \overline{\hat{\mathrm{N}}_1 + l_1, \hat{\mathrm{N}}_1 + \hat{\mathrm{N}}_3 - l_2}$ that is, really, reordered in an ascending order of their values the true output $y(k)$ with small portions l_1, l_2 of missing observations within it that belong, as we assume, to the left-hand and right-hand side sets of the data. Here arbitrary small enough integers $l_1, l_2 > 0$.

To calculate an auxiliary signal $\hat{x}(k)$ (the estimate of unmeasurable $x(k)$) $\forall\, k \in \overline{1, \mathrm{N}}$ one could approximate the model of the linear part of the Wiener system (7.1) by the finite impulse response (FIR) system of the form [11]

$$\tilde{y}(\tilde{k}) = \tilde{h}_0\tilde{u}(\tilde{k}) + \tilde{h}_1\tilde{u}(\tilde{k}-1) + \cdots + \tilde{h}_{\nu-1}\tilde{u}(\tilde{k}-\nu+1) + \tilde{h}_\nu\tilde{u}(\tilde{k}-\nu) + \tilde{e}(\tilde{k}) \tag{7.8}$$

$\forall\, \tilde{k} \in \overline{\hat{\mathrm{N}}_1 + l_1, \hat{\mathrm{N}}_1 + \hat{\mathrm{N}}_3 - l_2}$, or the expression in a matrix form

$$\tilde{\mathbf{Y}} = \tilde{\mathbf{\Lambda}}\tilde{\mathbf{h}}; \tag{7.9}$$

$$\tilde{\mathbf{Y}} = \left(\tilde{y}(\hat{\mathrm{N}}_1 + l_1), \tilde{y}(\hat{\mathrm{N}}_1 + l_1 + 1), \ldots, \tilde{y}(\hat{\mathrm{N}}_1 + \hat{\mathrm{N}}_3 - l_2 - 1), \tilde{y}(\hat{\mathrm{N}}_1 + \hat{\mathrm{N}}_3 - l_2)\right)^T \tag{7.10}$$

is the $(\hat{\mathrm{N}}_3 - l_1 - l_2 + 1) \times 1$ vector of the middle data set of $\tilde{y}(\tilde{k})$ without the small portions of observations from left-hand $\hat{\mathrm{N}}_1$ and from the right-hand side $\hat{\mathrm{N}}_2$ sets of the data, reordered in an ascending order, respectively; ν is the order of the FIR filter that can be arbitrarily large but fixed,

$$\tilde{\mathbf{\Lambda}} = \begin{bmatrix} \tilde{u}(\hat{\mathrm{N}}_1 + l_1) & \ldots & \tilde{u}(\hat{\mathrm{N}}_1 + l_1 - \nu + 1) & \tilde{u}(\hat{\mathrm{N}}_1 + l_1 - \nu) \\ \tilde{u}(\hat{\mathrm{N}}_1 + l_1 + 1) & \ldots & \tilde{u}(\hat{\mathrm{N}}_1 + l_1 - \nu + 2) & \tilde{u}(\hat{\mathrm{N}}_1 + l_1 - \nu + 1) \\ \vdots & \vdots & \vdots & \vdots \\ \tilde{u}(\hat{\mathrm{N}}_1 + \hat{\mathrm{N}}_3 - l_2) & \ldots & \tilde{u}(\hat{\mathrm{N}}_1 + \hat{\mathrm{N}}_3 - l_2 - 1 - \nu) & \tilde{u}(\hat{\mathrm{N}}_1 + \hat{\mathrm{N}}_3 - l_2 - \nu) \end{bmatrix} \tag{7.11}$$

is the full rank $(\hat{\mathrm{N}}_3 - l_1 - l_2 + 1) \times (\nu + 1)$ regression matrix, consisting only of observations of the non-noisy input $\tilde{u}(\tilde{k})$. Here

$$\tilde{\mathbf{h}}^T = \left(\tilde{h}_0, \tilde{h}_1, \ldots, \tilde{h}_\nu\right) \tag{7.12}$$

is a FIR of a linear block of Wiener system (where $\tilde{\mathbf{h}}$ is $(\nu + 1)$ vector of unknown parameters, mathematically), $\tilde{u}(\tilde{k})$ are observations of $u(k)$ associated with their own $\tilde{y}(\tilde{k})$, $\tilde{e}(\tilde{k})$ is a residual. Note that Eq. (7.8) is known as convolution sum which issue is a third signal based on two ones.

In this case, the dependence of some regressors on the process output will be facilitated, and the assumption of the ordinary LS that the regressors depend only on the non-noisy input signal, will be satisfied [9]. This is the main consequence of replacing the initial transfer function $G(q^{-1}, \mathbf{\Theta})$ of the linear part of the Wiener system by the FIR filter. In such a case, one avoids the influence of some missing regressors, appearing in the regression matrix $\mathbf{\Lambda}$, when the infinite impulse response (IIR) system is used. Then, the parametric estimation technique, based on ordinary LS, could be applied in the estimation of parameters (7.12) of the given FIR system (7.9), using the reordered observations of the middle data-set $\tilde{y}(k)\ \forall\ k \in \overline{\mathrm{N}_1 + l_1, \mathrm{N}_1 + \mathrm{N}_3 - l_2}$, because the rearrangement of observations does not influence the accuracy of LS estimates to be calculated. Note that, using the desription proposed in **V***örös* [17, 18] all the model parameters, including that of the nonlinearity, appear in the model explicitly. It allows, with the parameters of the linear block to estimate jointly the coefficients of the nonlinear function. In [2] the FIR mathematical model is used for the identification and predictive control of a distillation column.

To estimate the FIR parameters vector $\tilde{\mathbf{h}}$, one can use the ordinary LS expression in a matrix form [10]:

$$\hat{\tilde{\mathbf{h}}} = \left(\tilde{\mathbf{\Lambda}}^T \tilde{\mathbf{\Lambda}}\right)^{-1} \tilde{\mathbf{\Lambda}}^T \tilde{\mathbf{Y}}, \tag{7.13}$$

where

$$\hat{\tilde{\mathbf{h}}}^T = \left(\hat{\tilde{h}}_0, \hat{\tilde{h}}_1, \ldots, \hat{\tilde{h}}_\nu\right) \tag{7.14}$$

is a $(\nu + 1) \times 1$ vector of the estimates of parameters (7.12).

Let us formulate now the Theorem that determines the adequacy among distinct FIR parameter estimates accordingly their covariance matrices.

Theorem 1 *Assume that there exists the system of linear equations of the form*

$$\mathbf{Y} = \mathbf{\Pi}\beta + \mathbf{e}, \tag{7.15}$$

where $\mathbf{Y}$ *is a vector of observation samples,* $\mathbf{\Pi}$ *is a regression matrix,* β *is a vector of parameters, and* $\mathbf{e}$ *is a vector of residuals.*

Then the covariance matrix

$$cov\{\hat{\beta}\} = \sigma^2(\mathbf{\Pi}^T\mathbf{\Pi})^{-1} \tag{7.16}$$

and any other covariance matrix

$$cov\{\hat{\beta}_*\} = \sigma^2(\mathbf{\Pi}_*^T\mathbf{\Pi}_*)^{-1} \tag{7.17}$$

of parameter estimates $\hat{\beta}$ *and* $\hat{\beta}_*$*, respectively, are equivalent, i.e.*

$$cov\{\hat{\beta}\} \equiv cov\{\hat{\beta}_*\}, \tag{7.18}$$

if the matrix $\mathbf{\Pi}_*$ *and vector* $\mathbf{Y}_*$ *are obtained by interchanging rows in the initial matrix* $\mathbf{\Pi}$ *and in the initial vector* $\mathbf{Y}$ *in the same way. In Eqs.* (7.16) *and* (7.17) σ *is the variance of residuals.*

The proof follows from the aproved statement: the interchanging equations in the system (7.15) *transforms the system to an equivalent one [6].*

Theorem 2 *Let us assume that the* **Theorem** *1 holds. Then, the estimate*

$$\hat{\beta} = \left(\mathbf{\Pi}^T\mathbf{\Pi}\right)^{-1}\mathbf{\Pi}^T\mathbf{Y}, \tag{7.19}$$

and the estimate

$$\hat{\beta}_* = \left(\mathbf{\Pi}_*^T\mathbf{\Pi}_*\right)^{-1}\mathbf{\Pi}_*^T\mathbf{Y}_* \tag{7.20}$$

are coincidental if matrix $\mathbf{\Pi}_*$ *and vector* $\mathbf{Y}_*$ *are obtained by interchanging the rows in matrix* $\mathbf{\Pi}$ *and vector* $\mathbf{Y}$ *in the same way.*

Theorem 3 *The covariance matrix* (7.16) *and the estimate* (7.19) *are invariable to the rows interchanging in the initial vector* $\mathbf{Y}$ *and in the initial matrix* $\mathbf{\Pi}$*, respectively, if the rows interchanging in* $\mathbf{Y}$ *and in* $\mathbf{\Pi}$ *is made in the same way.*

Theorem 1 follows from **Theorems** 1 and 2.

Note that all proofs based on the deterministic regression matrix are valid here, too.

The estimate $\hat{x}(k)$ of the intermediate signal $x(k)$ could be determined using Eq. (7.8) where, instead of the true vector **h** components, their estimates are substituted, i.e.,

$$\hat{x}(m) = \hat{h}_0\tilde{u}(m) + \hat{h}_1\tilde{u}(m-1) + \cdots + \hat{h}_{m-1}\tilde{u}(m-l+1) + \hat{h}_l\tilde{u}(m-l). \quad (7.21)$$

The result of this step is the auxiliary signal $\hat{x}(k)$ that is a reconstructed version of the intermediate signal $x(k)$. It will be used at the next step to calculate the estimate of parameters $\boldsymbol{\Theta}$ of the transfer function $G(q^{-1}, \boldsymbol{\Theta})$ according to

$$\hat{\boldsymbol{\Theta}} = \left(\tilde{\boldsymbol{\Omega}}^T\tilde{\boldsymbol{\Omega}}\right)^{-1}\tilde{\boldsymbol{\Omega}}^T\hat{\mathbf{X}}. \quad (7.22)$$

Here

$$\hat{\boldsymbol{\Theta}}^T = (\hat{\mathbf{b}}, \hat{\mathbf{a}})^T, \hat{\mathbf{b}}^T = (\hat{b}_1, \ldots, \hat{b}_{\mathrm{n}}), \hat{\mathbf{a}}^T = \left(\hat{a}_1, \ldots, \hat{a}_{\mathrm{n}}\right) \quad (7.23)$$

are $2\mathrm{n} \times 1, n \times 1, n \times 1$ vectors of the parameter estimates, respectively,

$$\tilde{\boldsymbol{\Omega}} = \begin{bmatrix} \tilde{u}(0) & \ldots\tilde{u}(1-n) & \hat{x}(0) & \ldots\hat{x}(1-n) \\ \tilde{u}(1) & \ldots\tilde{u}(2-n) & \hat{x}(1) & \ldots\hat{x}(2-n) \\ \vdots & \vdots & \vdots & \vdots \\ \tilde{u}(k-1) & \ldots\tilde{u}(k-n) & \hat{x}(k-1) & \ldots\hat{x}(k-n) \end{bmatrix} \quad (7.24)$$

is the $k \times 2n$ matrix, consisting of reordered input $\tilde{u}(k)$ and reordered output $\hat{x}(k)$, values and

$\hat{\mathbf{X}} = \left(\hat{x}(1), \hat{x}(2), ..., \hat{x}(k-1), \hat{x}(k)\right)^T$
is the $k \times 1$ vector, consisting of the observation samples $\hat{x}(k)$. Note that coefficients $(\hat{b}_1, \ldots, \hat{b}_{\mathrm{n}})$ and $(\hat{a}_1, \ldots, \hat{a}_{\mathrm{n}})$ are needed in calculations of MVC law while the controller's training.

Estimates of the parameters c_0, d_0 and c_1, d_1 are calculated by the ordinary LS, too. In such a case, the criteria

$$I(c_0, c_1) = \frac{1}{2(\mathrm{N}_1 - l_3)} \sum_{i=1}^{\mathrm{N}_1 - l_3} \left[\tilde{y}(i) - c_0 - c_1\hat{x}(i)\right]^2, \quad (7.25)$$

$$I(d_0, d_1) = \frac{1}{2(\mathrm{N}_3 + l_4)} \sum_{j=\mathrm{N}_3 + l_4}^{\mathrm{N}} \left[\tilde{y}(j) - d_0 - d_1\hat{x}(j)\right]^2, \quad (7.26)$$

to be minimized accordingly of parameters c_0, c_1 and d_0, d_1, respectively, using side-set data particles of $\tilde{y}(k)$ and associated observations of the auxiliary signal $\hat{x}(k)$ in left side of Eq. (7.21).

Here $\hat{x}(k)$ are the observations of the signal $\hat{x}(k)$ in Eq.(7.21) that are rearranged in accordance with $\tilde{y}(k)$, arbitrary integers $l_3, l_4 > 0$.

The estimates of parameters c_1, d_1 and c_0, d_0 are calculated according to [8] as follows:

$$\hat{c}_1 = \frac{\sum_{i=1}^{N_1-l_3} \tilde{y}(i)\hat{x}(i)}{\sum_{i=1}^{N_1-l_3} \tilde{\hat{x}}^2(i)}, \qquad \hat{d}_1 = \frac{\sum_{j=1}^{N_3-l_4} \tilde{y}(j)\hat{x}(j)}{\sum_{j=1}^{N_3-l_4} \tilde{\hat{x}}^2(j)}, \tag{7.27}$$

$$\hat{c}_0 = \frac{\sum_{i=1}^{N_1-l_3}[\tilde{y}(i) - \hat{c}_1\hat{x}(i)]}{N_1 - l_3}, \qquad \hat{d}_0 = \frac{\sum_{j=1}^{N_3-l_4}[\tilde{y}(j) - \hat{d}_1\hat{x}(j)]}{N_3 - l_4}, \tag{7.28}$$

respectively, but using side-sets data particles of $\tilde{y}(k)$ and associated observations of the auxiliary signal $\hat{x}(k)$, that are reordered in accordance with $\tilde{y}(k)$.

At last, the estimates of the threshold a on the right-hand side and left-hand side sets are found according to

$$\hat{a} = \hat{d}_0/(1 + \hat{d}_1), \hat{a} = -\hat{c}_0/(1 + \hat{c}_1), \tag{7.29}$$

It should be noted that the simple input and associated output data reordering with a following reconstruction of an intermediate signal $x(k)$ that is really unknown, allow us in the training phase to turn the nonlinear problem of parametric estimation of Wiener systems to a linear one, where linear estimators based on the ordinary LS, could be applicable for each reordered data set. In such a case, all observations of the input are available for processing. Therefore the FIR model could be used to replace the initial transfer function $G(q^{-1}, \Theta)$ of the linear part of the Wiener system. Thus, the dependence of some regressors on the process output will be diminished, and the assumption that the regressors depend only on the non-noisy input signal will be satisfied. Besides, by applying the FIR model for the parametric identification of the Wiener system one avoids the influence of some missing regressors, appearing in the regression matrix. This opportunity does not take place in the Hammerstein system due to the presence of some cut off or missing observation samples in the input sequence.

The estimates consistency problem is studied in detail by Ljung [7] and developed by Hagenblad [4]. The development is rather general, and applies in particular to Wiener models. A brief outline her results, which are in a close contact with our ones is represented in Appendix.

7.3 Numerical Simulation for the Wiener System Having the Invertible Nonlinearity

The true intermediate signal $\{x(k)\}$ $k = \overline{1, N}$, of the Wiener system (Figs. 7.4b) and 7.5b) is given by (7.5) with additive process noise $\{v(k)\} \equiv 0$. The true output signal (Figs. 7.4c and 7.5c) is described by

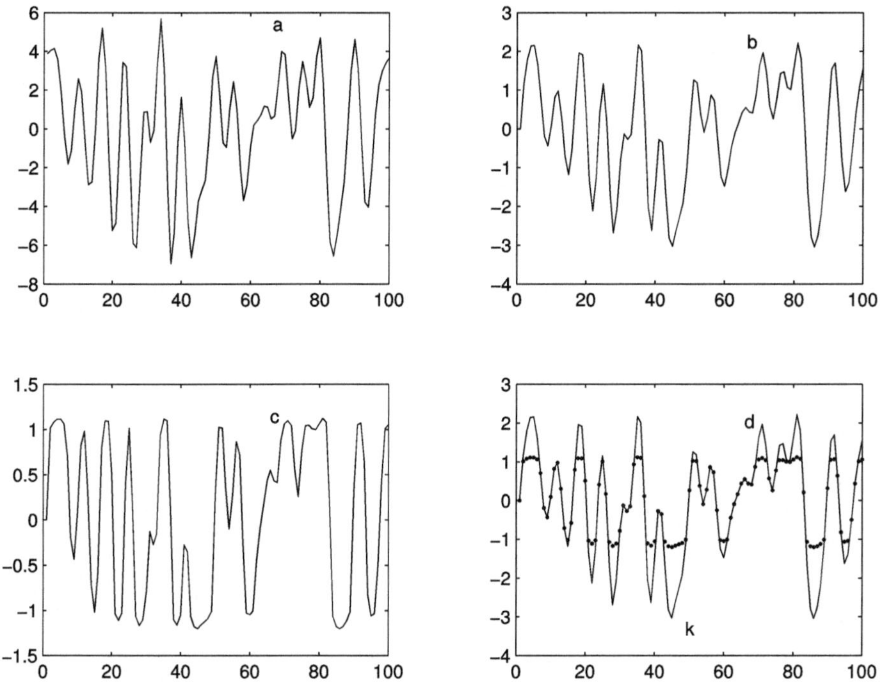

Fig. 7.4 The signals of the simulated LTI system with a PWL nonlinearity (7.30): input $u(k)$, calculated by (7.31) (**a**), intermediate signal $x(k)$ (**b**), output $y(k)$ (converted signal $x(k)$) (**c**), intermediate and output (dotted line) signals (**d**)

$$y(k) = \begin{cases} -0.9 + 0.1x(k) & \text{if } x(k) \le -1, \\ x(k) & \text{if } -1 < x(k) \le 1, \\ 0.9 + 0.1x(k) & \text{if } x(k) > 1 \end{cases} \tag{7.30}$$

with the sum of sinusoids (Fig. 7.4a)

$$u(k) = \frac{1}{20} \sum_{i=1}^{20} \sin(i\pi k/10 + \varphi_i) \tag{7.31}$$

and white Gaussian noise (Fig. 7.5a) as inputs or training signals to the linear block

$$G(q^{-1}, \boldsymbol{\Theta}) = \frac{b_1 q^{-1}}{1 + a_1 q^{-1}} = \frac{0.3q^{-1}}{1 - 0.5q^{-1}}. \tag{7.32}$$

In (7.31) random variables φ_k with a uniform distribution on $[0, 2\pi]$.

N = 100 samples have been generated in an unnoisy frame (Figs. 7.4–7.8 and 7.9). Then, the various dependences, including those between input-output signals, reordered in an ascending order, are obtained by Matlab. Results, shown on

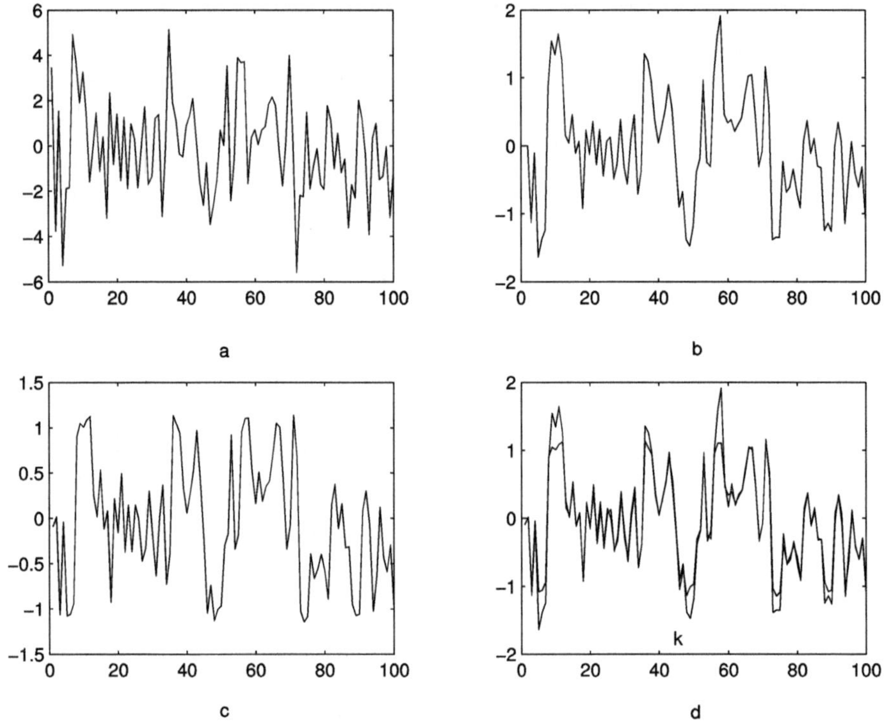

Fig. 7.5 The signals of the simulated PWA system with a PWL nonlinearity (7.30): input $u(k)$ is white Gaussian noise (**a**), intermediate signal $x(k)$ (**b**), output $y(k)$ (converted signal $x(k)$) (**c**), intermediate and output signals (**d**)

(Fig. 7.6 and 7.7) allow us to make conclusions, firstly, for invertible PWL nonlinearity (7.30), the dependence of reodered output $\mathtt{sort}(\{y(k)\})$ of which on reodered input $\mathtt{sort}(\{x(k)\})$ could be used as curve holding the right information on thresholds, slopes, etc., of PWL nonlinearity (Figs. 7.6 and 7.7, parts d). Remaining dependences are not so informative. Afterwards, the LS problem (7.13) was solved, using 50 and 70 rearranged observations of the output, respectively (Figs. 7.10c and 7.11c), excluding zeros. The whole number of FIR filter parameters $\nu = 14$ has been chosen based on the estimation results (Tables 7.1 and 7.2), obtained for different ν in the absence of process and measurement noises. The estimate $\hat{x}(k)$ of the intermediate signal $x(k)$ was reconstructed according to (7.21). The reconstructed versions of the intermediate signal $x(k)$ are shown in (Figs. 7.12a and 7.13a). The estimates $\hat{\mathbf{\Theta}}^T = (b_1, a_1)$ of parameters $\mathbf{\Theta}$ of the transfer function $G(q, \mathbf{\Theta})$ were calculated by Eq. (7.22), using the observations of the auxiliary signal $\hat{x}(k)$. Estimates $\hat{b}_1$ and $\hat{a}_1$ could be used to calculate the current value of control signal $u(k)$. The MV controller for the LTV system without any nonlinear distortion is

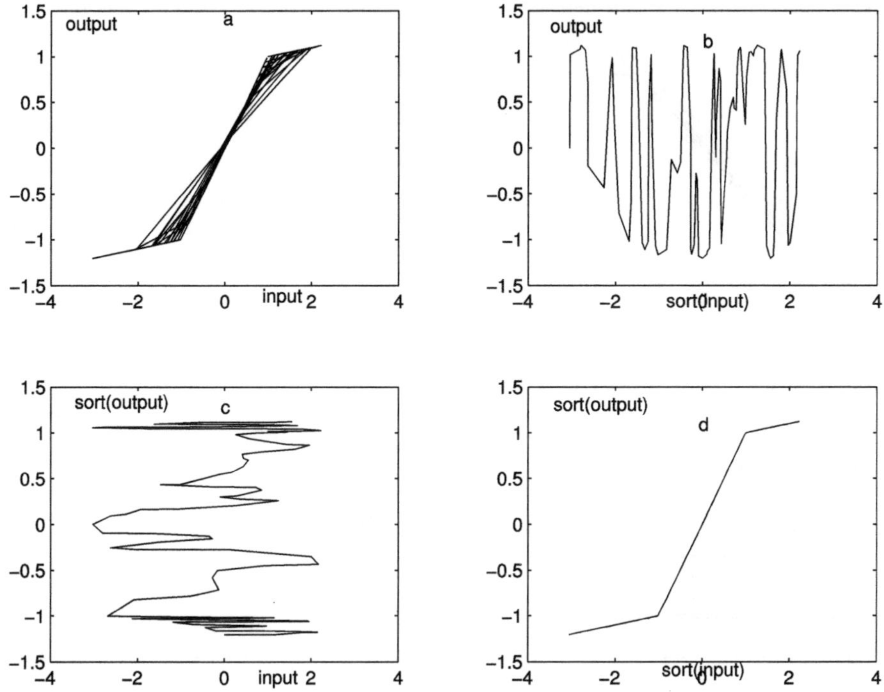

Fig. 7.6 Input-output dependences. Dep.-e of output signal $\{y(k)\}$ on the input $\{x(k)\}$ (**a**), dep.-e of $\{y(k)\}$ on reodered `sort`$(\{x(k)\})$ (**b**), dep.-e of `sort`$(\{y(k)\})$ on $x(t)$ (**c**), and dep.-e of `sort`$(\{y(k)\})$ on `sort`$(\{x(k)\})$ (**d**), where `sort` is Matlab routine. Input $\{x(k)\}$ is of the form (7.31)

$$u(k-1) = \frac{\hat{a}_1 x(k-1) + r(k)}{\hat{b}_1}, \tag{7.33}$$

Here $\hat{b}_1 \equiv \hat{b}_1(k)$, $\hat{a}_1(k) \equiv \hat{a}_1(k)$ are current values of the linear system (block) parameter estimates that are calculated at the k-th recursive iteration; $\hat{b}_1(k) \neq 0$. Afterwards, the estimate $\hat{x}_1(k)$ of the intermediate signal $x(k)$ was recalculated by

$$\hat{x}_1(k) = \hat{b}_1\, u(k-1) + \hat{a}_1 \hat{x}_1(k-1), \quad \forall \;\; k = \overline{2, 100}, \tag{7.34}$$

using $\hat{b}_1, \hat{a}_1$ and $\hat{x}_1(1) = 0$. In such a case, the estimates $\hat{b}_1, \hat{a}_1$ were equal to the true parameters: $b_1 = 0.3, a_1 = 0.5$. The reconstructed versions of the intermediate signal $x(k)$ are shown in (Figs. 7.12b and 7.13b).

Note that the accuracy of estimates of the intermediate signal, calculated by formulas (7.21) and (7.34), is more or less similar except for the first 15 observations, when the FIR model (7.21) was used. If $\hat{x}(k)$ has been obtained, then it is simple to separate different particles of observations that belong to the respective side-sets. The estimates of parameters c_1, d_1 and c_0, d_0 are calculated according to formu-

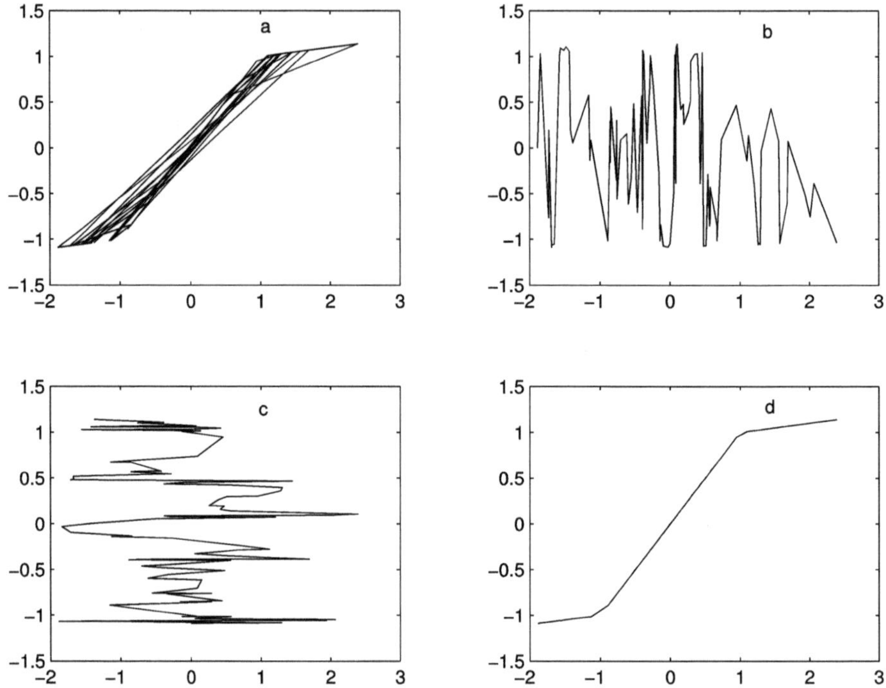

Fig. 7.7 The values, notation and markings are the same as in Fig. 7.6, except that $\{x(k)\}$ is white Gaussian noise realization

las (7.27) and (7.28), respectively. In such a case, the rearranged observations of $\hat{x}(k)$ and $y(k)$ were substituted in formulas (7.27) and the estimates of c_1 and d_1 were determined: $\hat{c}_1 = \hat{d}_1 = 0.1$. The estimates $\hat{c}_0$ and $\hat{d}_0$ were calculated by (7.28). Their values are also coincidental with the values of true coefficients: $\hat{c}_0 = -0.9$, while $\hat{d}_0 = 0.9$. It should be noted that $N_1 = 31$, $N_2 = 29$ for the periodical signal (7.31) (Fig. 7.4a) and $N_1 = 15$, $N_2 = 12$ for the Gaussian white noise (Fig. 7.5a) were used to calculate the estimates $\hat{c}_0, \hat{c}_1, \hat{d}_0, \hat{d}_1$, respectively. The estimates of the threshold were established by Eq. (7.29). The values of estimates $\hat{a}$ were equal to the true value a=1. In order to determine how realizations of different process- and measurement noises affect the accuracy of estimation of unknown parameters, we have used the Monte Carlo simulation with 10 data sets, each containing 100 pairs of input-output observations with additive process and measurement noises according to $x(k) = x(k)+\lambda_1 v(k)$, $y(k) = y(k)+\lambda_2 v(k)$ $\forall\, k = \overline{1, 1000}$, respectively. Here values λ_1, λ_2 were chosen so that SNRs (the square root of the ratio of signal and noise variances) were equal to 10 for the process noise $v(k) \forall k \in \overline{1, 1000}$, as well as for the measurement noise $n(k) \forall k \in \overline{1, 1000}$, and to 20 for measurement noise $n(k)$. The estimates $\bar{\hat{b}}_1(k)$, $\bar{\hat{b}}_2(k)$, and $\bar{\hat{a}}_1(k)$, $\bar{\hat{a}}_2(k)$ are:

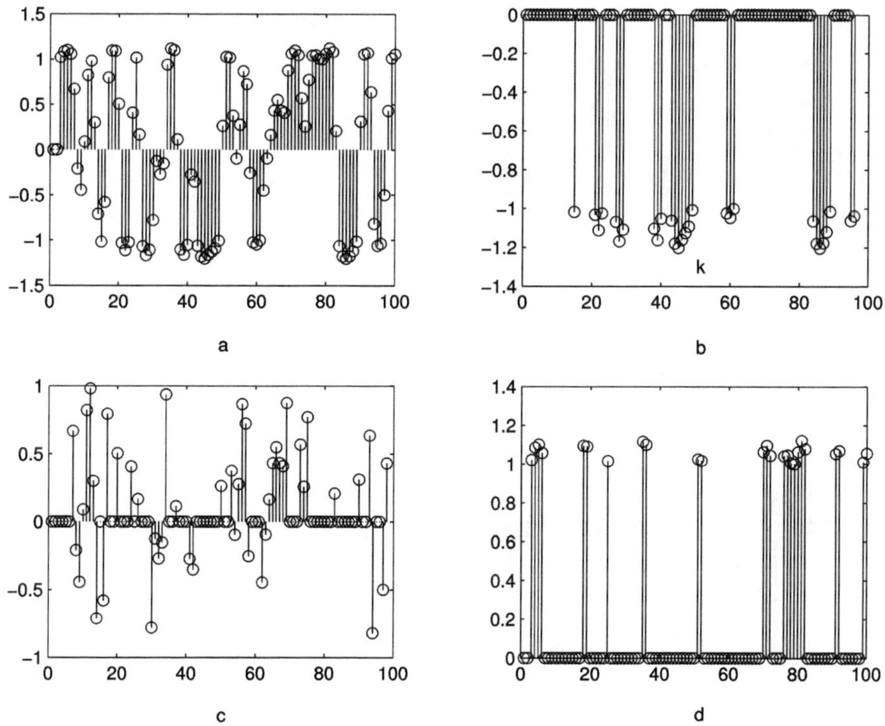

Fig. 7.8 Samples of $y(k)$ (**a**) (see Fig. 7.4c) and its data sets: left (**b**), middle (**c**), right (**d**) (here the observations, that belong to the other data set, are equal to zeros). Input $u(k)$ of the form (7.31) (see Fig. 7.4a)

$$\bar{\hat{b}}_1(k) = \frac{1}{10}\sum_{i=1}^{10} \hat{b}_0^{(i)}(k),\ \bar{\hat{b}}_2(k) = \frac{1}{10}\sum_{i=1}^{10} \hat{b}_1^{(i)}(k), \tag{7.35}$$

$$\bar{\hat{a}}_1(k) = \frac{1}{10}\sum_{i=1}^{10} \hat{a}_1^{(i)}(k),\ \bar{\hat{a}}_2(k) = \frac{1}{10}\sum_{i=1}^{10} \hat{a}_2^{(i)}(k), \tag{7.36}$$

and their confidence intervals $\triangle$:

$$\triangle_{\hat{b}_1}^{(s)} = \pm t_\alpha \frac{\hat{\sigma}_{\hat{b}_1^{(s)}}}{\sqrt{s}},\ \triangle_{\hat{b}_2}^{(s)} = \pm t_\alpha \frac{\hat{\sigma}_{\hat{b}_2^{(s)}}}{\sqrt{s}}, \tag{7.37}$$

$$\triangle_{\hat{a}_1}^{(s)} = \pm t_\alpha \frac{\hat{\sigma}_{\hat{a}_1^{(s)}}}{\sqrt{s}},\ \triangle_{\hat{a}_2}^{(s)} = \pm t_\alpha \frac{\hat{\sigma}_{\hat{a}_2^{(s)}}}{\sqrt{s}}, \tag{7.38}$$

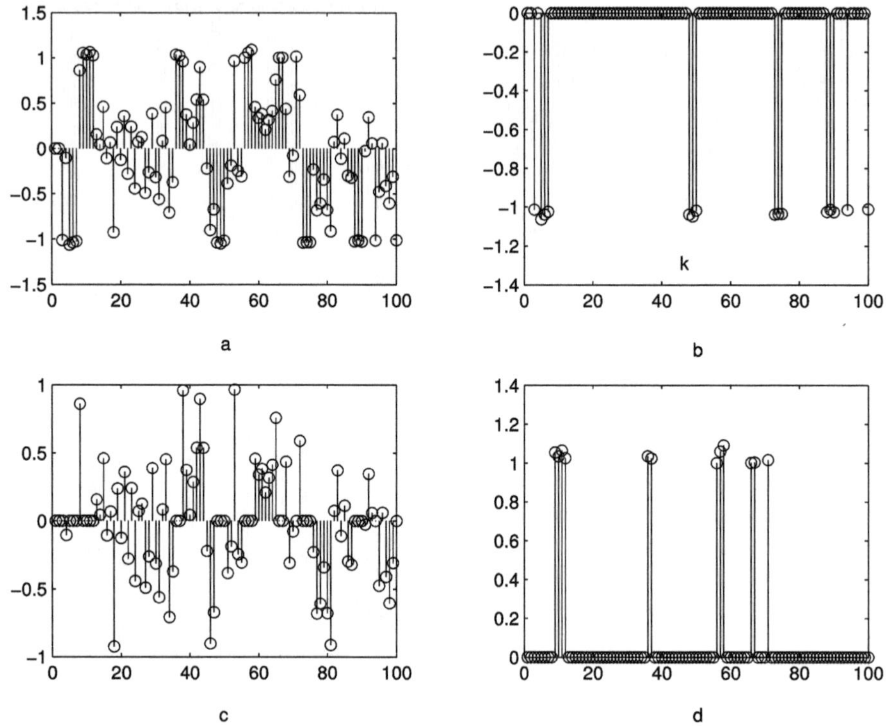

Fig. 7.9 Samples of signal $y(k)$ (**a**) (see Fig. 7.5c) and its data sets: left (**b**), middle (**c**), right (**d**). Input $u(k)$ is white Gaussian noise realization (see Fig. 7.5a)

averaged by 10 tracking experiments have been calculated after 10 experiments for each of the given input sequence.

Here $\hat{\sigma}_w^s$ is the estimate of variance σ_w^s at the calculated recursions s in total; α = 0.05 is the significance level, and t_α = 2.26 is the 100$(1 - \alpha)$ point of Student's distribution with $n = s - 1$ degrees of freedom.

10 experiments with the same realization of the process noise $v(k)$ and different realizations of the measurement noise $e(k)$ with different levels of its intensity have been carried out. The intensity of noises was assured by choosing respective signal-to-noise ratios SNR (the square root of the ratio of signal and noise variances). For the process noise, $\mathrm{SNR}^{\mathrm{v}}$ was equal to 100, and for the measurement noise, $\mathrm{SNR}^{\mathrm{e}}$: 1, 10, 100. As inputs (training signals) for all given nonlinearities the periodical signal (7.31) and white Gaussian noise with variance 1 were chosen. In each ith experiment the estimates of parameters were obtained. During the Monte Carlo simulation averaged values of estimates of the parameters and of the threshold and their confidence intervals were calculated. In Tables 7.3 and 7.4, for each input the averaged estimates of parameters and the threshold a of the simulated PWA system (Fig. 1.3) with the linear part (7.32) ($b_1 = 0.3$; $a_1 = -0.5$) and the PWL

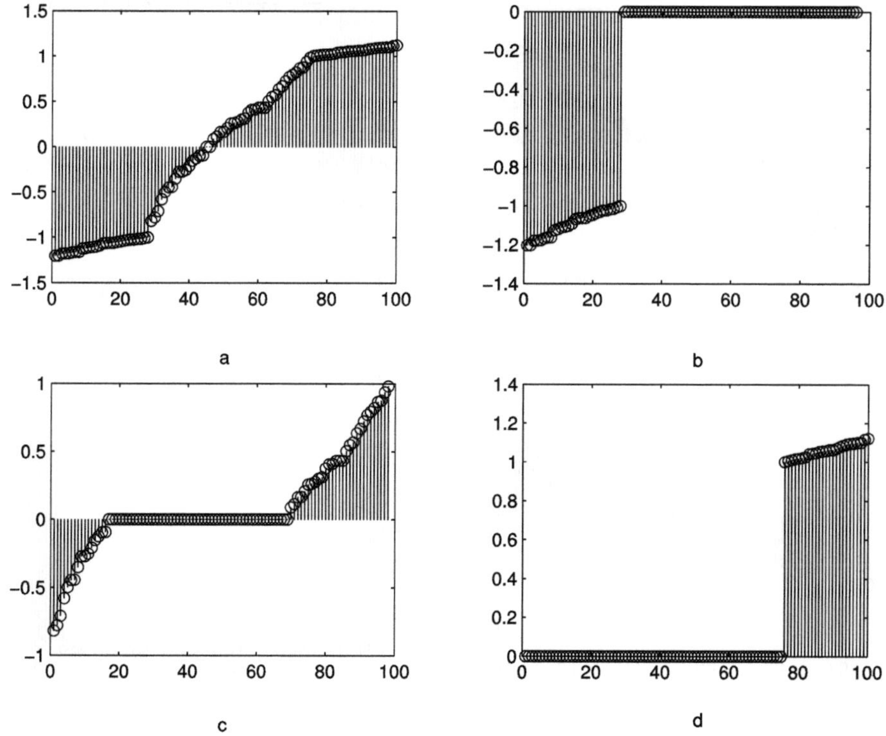

Fig. 7.10 The reordered in an ascending order of their values signal $y(k)$ (**a**) (see Figs. 7.4c and 7.8a) and its rearranged data sets: left (**b**), middle (**c**), right (**d**) (here the observations, that belong to the other data set, are equal to zeros). Input $u(k)$ of the form (7.31) (see Fig. 7.4a)

nonlinearity (7.30) ($c_0 = -0.9, c_1 = 0.1, d_0 = 0.9, d_1 = 0.1$) with their confidence intervals are given. Note that in each experiment here the value of SNR^v was fixed and was the same, while the values of SNR^e were varying due to different realizations of $e(k)$. The Monte Carlo simulation (Tables 7.3 and 7.4) implies that the accuracy of parametric identification of the PWL Wiener system depends strongly on the intensity of measurement noise.

7.4 Simulation for the Wiener System Having the Noninvertible Nonlinearity

It is emphasized in Hagenblad [4] that it may be possible to identify the linear system also if f is non-invertible. In such a case, the function f ought to be linear in a small region around the origin. Then it will be possible to identify the system if only the first two estimate consistency conditions (see Appendices C1 and C2), concerning

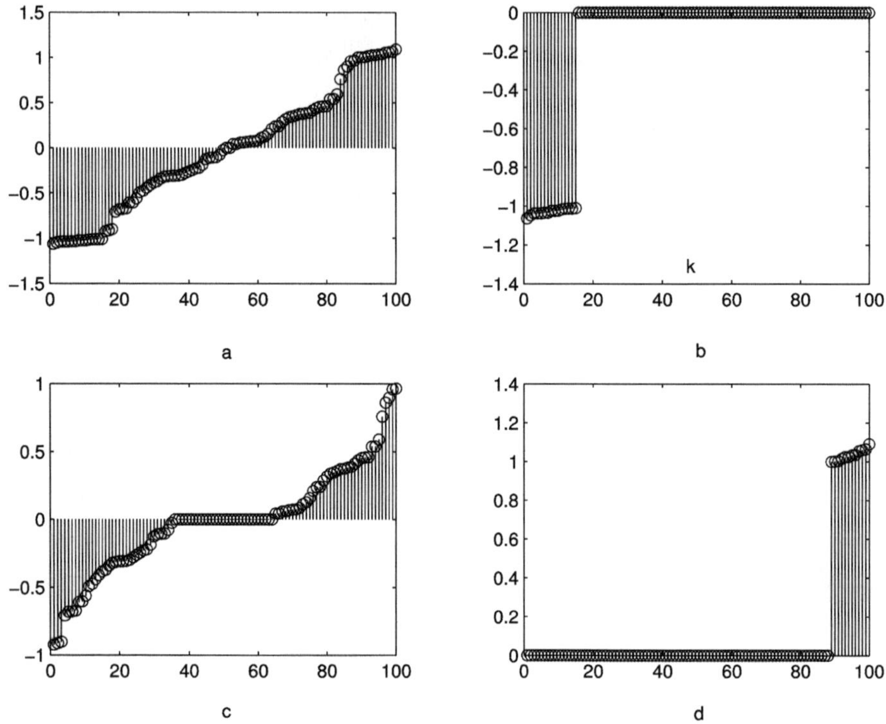

Fig. 7.11 The reordered in an ascending order of their values signal $y(k)$ (**a**) (see Figs. 7.5c and 7.9a) and its rearranged data sets: left (**b**), middle (**c**), right (**d**). Input $u(k)$ is white Gaussian noise realization (see Fig. 7.5a)

the global identifiability of the linear model structure and data sets informativiability, respectively, are satisfied. It will be shown here by Monte Carlo simulation that the method of data reordering allows us to identify not only the linear system as well as the non-invertible nonlinearity coefficients if the threshold d is known in advance [13].

The true intermediate signal $x(k)$ $k = \overline{1, N}$, of the PWA system (Figs. 7.14b and 7.15b) is given by (7.5) with additive process noise $\{v(k)\} \equiv 0$. The true output signal (Figs. 7.14c and 7.15c) is described by

$$y(k) = \begin{cases} -7.6 - 0.1x(k) & \text{if } x(k) \leq -7.5, \\ x(k) & \text{if } -7.5 < x(k) \leq 7.5, \\ 7.6 - 0.1x(k) & \text{if } x(k) > 7.5 \end{cases} \tag{7.39}$$

and by

$$y(k) = \begin{cases} -1.1 - 0.1x(k) & \text{if } x(k) \leq -1, \\ x(k) & \text{if } -1 < x(k) \leq 1, \\ 1.1 - 0.1x(k) & \text{if } x(k) > 1 \end{cases} \tag{7.40}$$

Table 7.1 The dependence of estimates of the parameters $b_1, a_1, c_0, c_1, d_0, d_1$, and thresholds a, $-a$ on the number of the FIR parameters ν. Input: the periodical signal (7.31)

Estimates	$\nu = 5$	$\nu = 10$	$\nu = 14$
$\hat{b}_1$	0.28	0.29	0.3
$\hat{a}_1$	−0.49	−0.5	−0.5
$\hat{c}_0$	−0.9	−0.89	−0.89
$\hat{c}_1$	0.1	0.1	0.16
$\hat{d}_0$	0.9	0.9	1.0
$\hat{d}_1$	0.09	0.1	0.1
$\hat{a}$	0.92	1	1
$-\hat{a}$	−0.95	−1	−1

Table 7.2 The values and notation are the same as in Table 7.1. Input—the Gaussian white noise

Estimates	$\nu = 5$	$\nu = 10$	$\nu = 14$
$\hat{b}_1$	0.29	0.3	0.3
$\hat{a}_1$	−0.49	−0.49	−0.5
$\hat{c}_0$	−0.89	−0.89	−0.9
$\hat{c}_1$	0.09	0.09	0.1
$\hat{d}_0$	0.89	0.9	0.9
$\hat{d}_1$	0.09	0.1	0.1
$\hat{a}$	0.96	0.98	1
$-\hat{a}$	−0.97	−0.99	−1

with the sum of sinusoids (7.31) (Fig. 7.14a) and white Gaussian noise with variance 1 (Fig. 7.15a) as inputs to the linear block (7.32). Both signals can be treated as the training ones. Characteristics between input-output signals, reordered in an ascending order, are obtained by Matlab programme. Results, shown on (Figs. 7.16 and 7.17) let us to notice that for non-invertible PWL nonlinearity the characteristic of dependency of reodered output `sort`$(\{y(k)\})$ on reodered input `sort`$(\{x(k)\})$ is not informative. It even takes us a wrong information on the mould of PWL nonlinearity, e.g. the curve instead of negative slopes displays positive ones (Figs. 7.16 and 7.17, parts d). Nevertheless, segment breakpoints could be determined, using above mentioned dependency, accurately enough. Remaining dependences are not informative except of characteristics given in (Figs. 7.16 and 7.17, parts a). These characteristics have some information on thresholds and slopes of nonlinearity. Such an information could be used as initial by solving the identification problem.

Here $b_1 = 1$, $a_1 = -0.7$; in (7.31) the stochastic variables ϕ_k with a uniform distribution on $[0, 2\pi]$ were chosen [4]. First of all, N = 100 data points have been generated without additive process and measurement noises (Figs. 7.18 and 7.19). Afterwards, the LS problem (7.13) was solved, using 38 and 36 rearranged observations of the output, respectively (Figs. 7.20c and 7.21c), excluding zeros. The whole

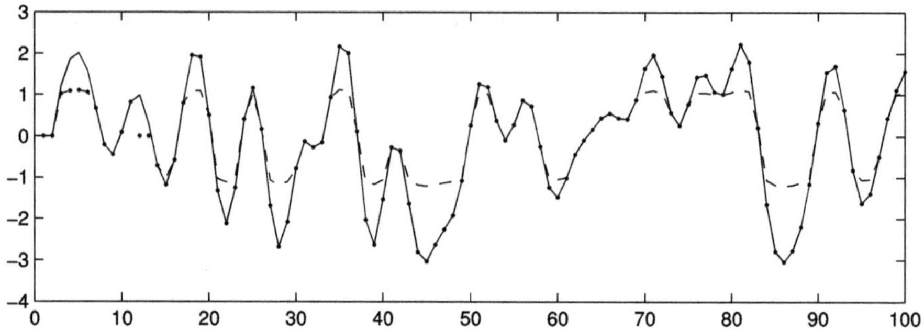

a

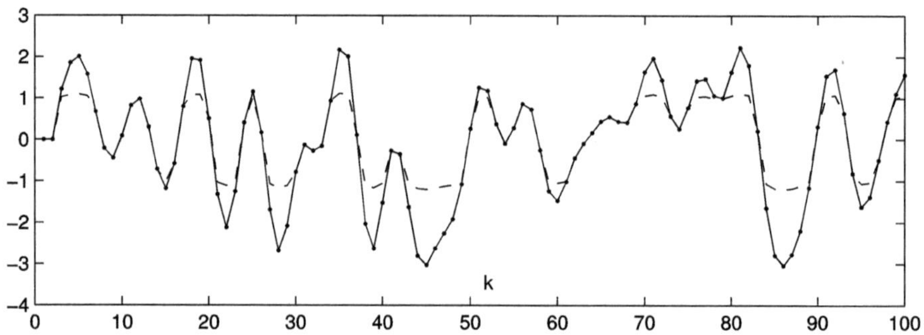

b

Fig. 7.12 The intermediate signal $x(k)$ (continuous line), the output signal $y(k)$ (dashed line), the reconstructed versions of $x(k)$ (dotted line), calculated using Eq. (7.21) (**a**) and Eq. (7.34) (**b**). Training signal is of the form (7.31). Continuous and dotted lines are coincidental

number of FIR filter parameters $\nu = 14$ has been chosen based on the estimation results, shown in Tables 7.1 and 7.2, obtained for different ν in the absence of process and measurement noises in Pupeikis 2005. The estimate $\hat{x}(k)$ of the intermediate signal $x(k)$ was reconstructed according to (7.21), replacing unknown true values of parameters by their estimates. The reconstructed versions of the intermediate signal $x(k)$ are shown in (Figs. 7.22a and 7.23a). The estimates $\hat{\mathbf{\Theta}}^T = (b_1, a_1)$ of parameters $\mathbf{\Theta}$ of the transfer function $G(q, \mathbf{\Theta})$ were calculated by common Eq. (7.22), using the observations of the auxiliary signal $\hat{x}(k)$. Afterwards, the estimate $\hat{x}_1(k)$ of the intermediate signal $x(k)$ was recalculated by (7.34) using $\hat{b}_1$, $\hat{a}_1$ and $\hat{x}_1(1) = 0$. The reconstructed versions of the intermediate signal $x(k)$, calculated by Eq. (7.34) are shown in (Figs. 7.22b and 7.23b).

The accuracy of estimates of the intermediate signal, calculated by formulas (7.21) and (7.34), is more or less similar except for the first 15 observations, when the FIR model (7.21) was used. It should be noted that N_1=14, N_2=8 for the periodical signal (7.31) (Fig. 7.14a), and N_1=32, N_2=28 for the Gaussian white noise (Fig. 7.15a). We have used the Monte Carlo simulation with 10 data sets, each containing 100 pairs

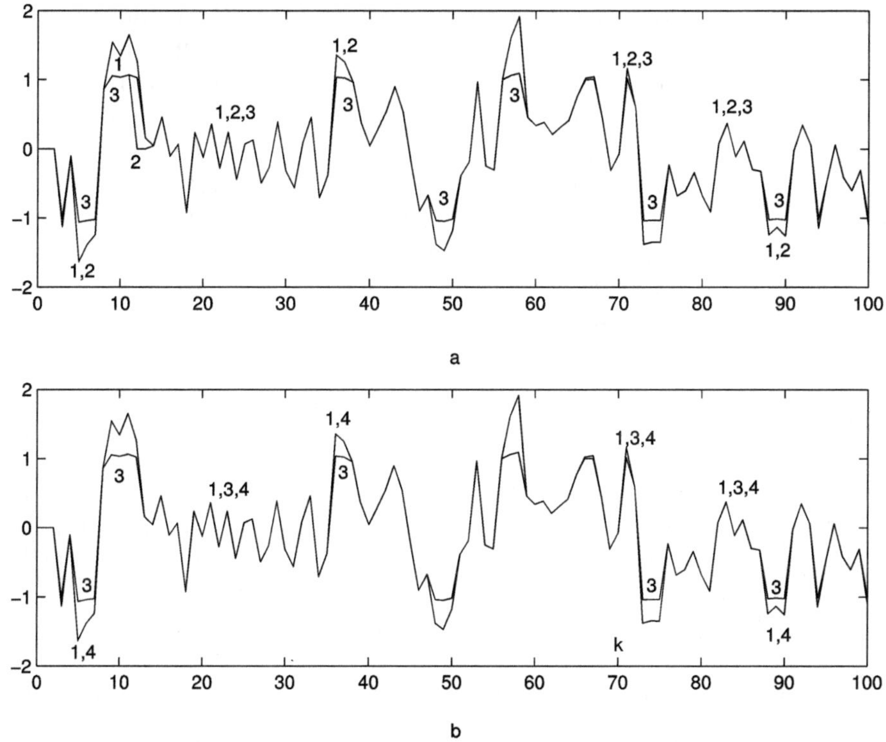

Fig. 7.13 The intermediate signal $x(k)$ (curve 1), the output signal $y(k)$ (curve 3), the reconstructed versions of $x(k)$ (curves 2, 4), calculated using Eq. (7.21) (**a**) and Eq. (7.34) (**b**), respectively. Training signal is white Gaussian noise realization

Table 7.3 Averaged estimates of the parameters $b_1, a_1, c_0, c_1, d_0, d_1$, and thresholds $a, -a$ with their confidence intervals. Input: the periodical signal (7.31). $\mathrm{SNR}^v = 100$

Estimates	$\mathrm{SNR}^e = 1$	$\mathrm{SNR}^e = 10$	$\mathrm{SNR}^e = 100$
$\hat{b}_1$	0.28 ± 0.07	0.3 ± 0.00	0.3 ± 0.00
$\hat{a}_1$	-0.52 ± 0.06	-0.5 ± 0.00	-0.5 ± 0.00
$\hat{c}_0$	-0.9 ± 0.00	-0.89 ± 0.04	-0.85 ± 0.31
$\hat{c}_1$	0.1 ± 0.00	0.1 ± 0.02	0.16 ± 0.21
$\hat{d}_0$	0.9 ± 0.0	0.89 ± 0.04	1 ± 0.5
$\hat{d}_1$	0.1 ± 0.0	0.1 ± 0.02	0.04 ± 0.4
$\hat{a}$	1 ± 0.00	1 ± 0.02	0.97 ± 0.25
$-\hat{a}$	-1 ± 0.00	-1 ± 0.02	-1 ± 0.22

Table 7.4 The values and notation are the same as in Table 7.3. Input—the Gaussian white noise

Estimates	$\mathrm{SNR}^e = 1$	$\mathrm{SNR}^e = 10$	$\mathrm{SNR}^e = 100$
$\hat{b}_1$	0.31 ± 0.04	0.3 ± 0.00	0.3 ± 0.00
$\hat{a}_1$	-0.39 ± 0.08	-0.5 ± 0.01	-0.5 ± 0.00
$\hat{c}_0$	-0.89 ± 0.00	-0.84 ± 0.06	-0.5 ± 0.86
$\hat{c}_1$	0.1 ± 0.00	0.1 ± 0.05	0.42 ± 0.72
$\hat{d}_0$	0.9 ± 0.00	0.91 ± 0.06	1.07 ± 0.51
$\hat{d}_1$	0.1 ± 0.00	0.15 ± 0.05	0.04 ± 0.44
$\hat{a}$	1 ± 0.0	0.99 ± 0.02	1.03 ± 0.4
$-\hat{a}$	-1 ± 0.00	-1 ± 0.02	-1.21 ± 0.19

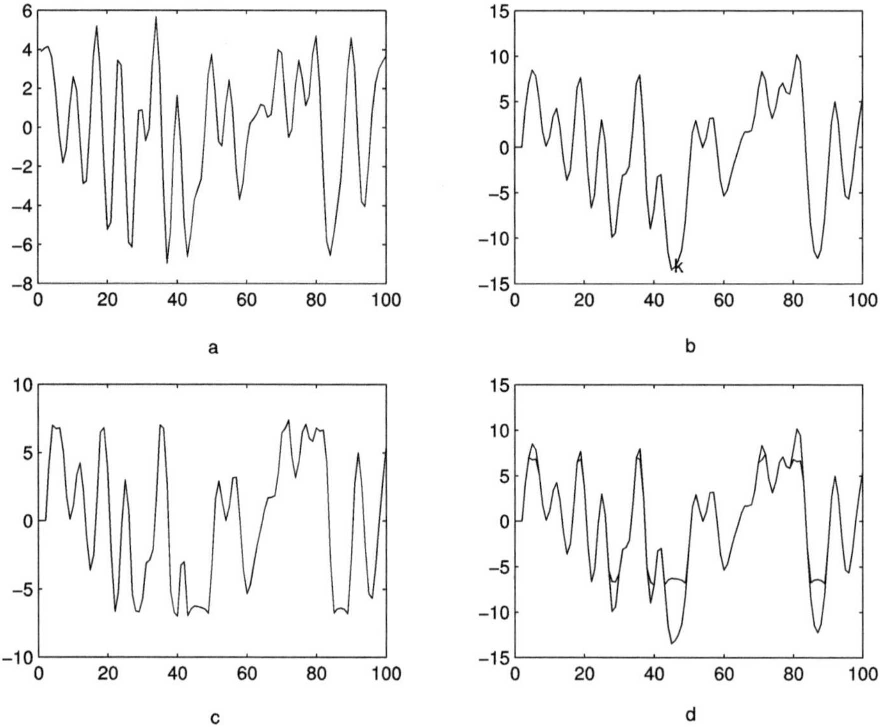

Fig. 7.14 The signals of the simulated PWA system with a piecewise nonlinearity (7.39): input $u(k)$, calculated by (7.31) (**a**), intermediate signal $x(k)$ (**b**), output $y(k)$ (**c**), intermediate and output (dotted line) signals (**d**)

of input-output observations. 10 experiments with the same realization of the process noise $v(k)$ and different realizations of the measurement noise $e(k)$ with different levels of its intensity have been carried out. The intensity of noises was assured by choosing respective signal-to-noise ratios SNR (the square root of the ratio of sig-

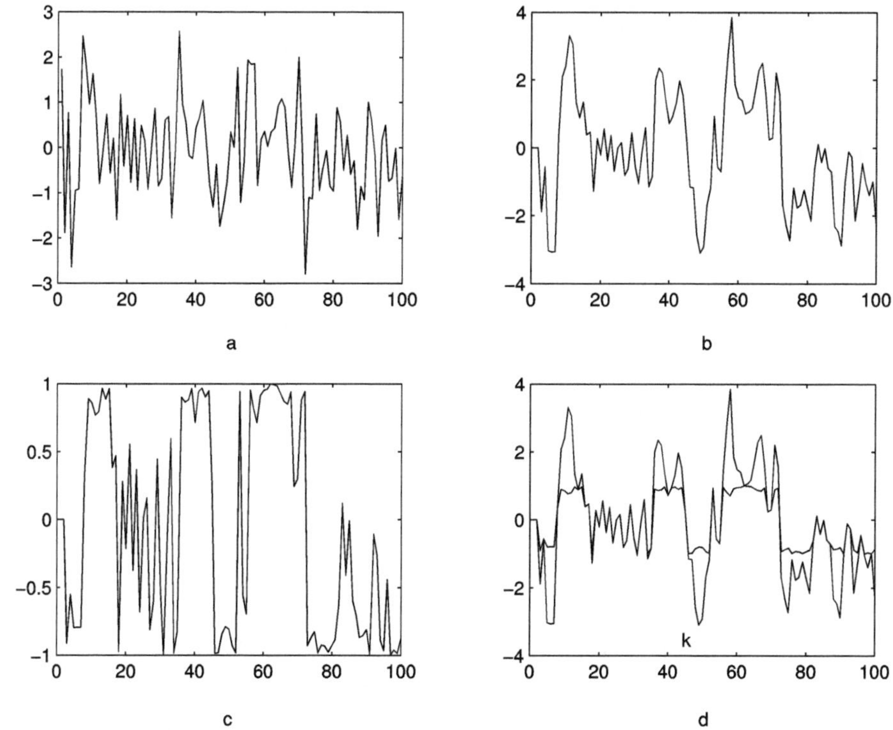

Fig. 7.15 The signals of the simulated PWA system with a piecewise nonlinearity (7.40): input $u(k)$ is white Gaussian noise (**a**), intermediate signal $x(k)$ (**b**), output $y(k)$ (**c**), intermediate and output signals (**d**)

nal and noise variances). For the process noise, $\mathrm{SNR}^{\mathrm{v}}$ was equal to 10, and for the measurement noise, $\mathrm{SNR}^{\mathrm{e}}$: 1, 10, 100. As training signals for the parameter estimation flowcharts with both given nonlinearities (7.39) and (7.40) the periodical signal (7.31) and white Gaussian noise with variance 1 were chosen, as usuall. In each ith experiment the estimates of parameters were calculated. During the Monte Carlo simulation averaged values of estimates of the parameters and of the threshold and their confidence intervals were calculated. In Tables 7.5 and 7.6, for each input the averaged estimates of parameters and the threshold a of the simulated PWA system (Fig. 1.3) with the linear part (7.32) ($b_1 = 1; a_1 = -0.7$) and the piecewise nonlinearities (7.39) and (7.40) having ($c_0 = -1.1, c_1 = 0.1, d_0 = 1.1, d_1 = 0.1$), and ($c_0 = -7.6, c_1 = 0.1, d_0 = 7.6, d_1 = 0.1$), respectively, with their confidence intervals are presented. Note that in each experiment the value of SNR^{v} was fixed and was the same, while the values of SNR^{e} were varying due to distinct realizations of $e(k)$. The Monte Carlo simulation (Tables 7.5 and 7.6) implies that the accuracy of parametric identification of the PWA system strongly depends on the intensity of measurement noise.

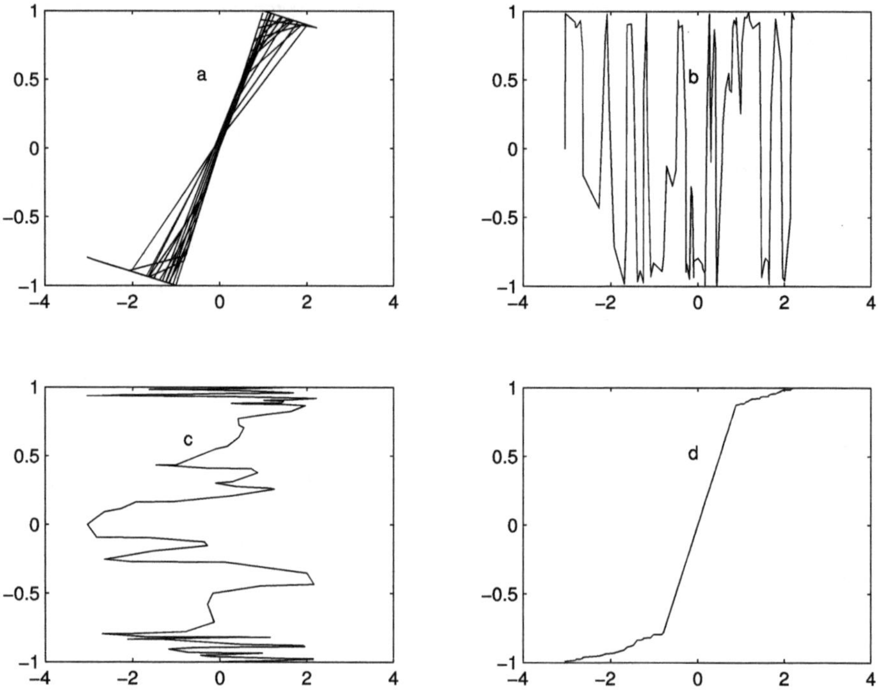

Fig. 7.16 Dependences, values, notations, and markings are the same as in Fig. 7.6.Input $x(k)$ is of the form (7.31)

7.5 Data Reordering for Hammerstein Systems

A special class of nonlinear systems applied in engineering is Hammerstein systems with hard input nonlinearities [1]. They, usually, consist of a static input nonlinearity and a linear dynamic system that are coupled together. Ordinary examples of hard nonlinearities are the saturation, preload, relay, deadzone, hysteresis-relay, and hysteresis nonlinearities [3]. However, the assumptions that the nonlinearity is invertible or linear in a small region around the origin are not satisfied for most hard nonlinearities, because they cannot be described by polynomials and are non-invertible in general. On the other hand, the Hammerstein systems are common in nonlinear control applications, e.g. see [1, 3, 16]. Frequently as an input nonlinearity the PWL nonlinearity (Eq. (7.6) and Fig. 7.1) is used here, too. Assuming the nonlinearity to be PWL one, we could let the nonlinear part of the Hammerstein system be represented by different regression functions with some parameters, that are unknown beforehand. In such a case, observations of the input of a Hammerstein system could be partitioned into distinct data sets according to different descriptions [12]. However, the boundaries of sets of observations depend on the value of the unknown threshold d—observations are divided into regimes subject to whether the

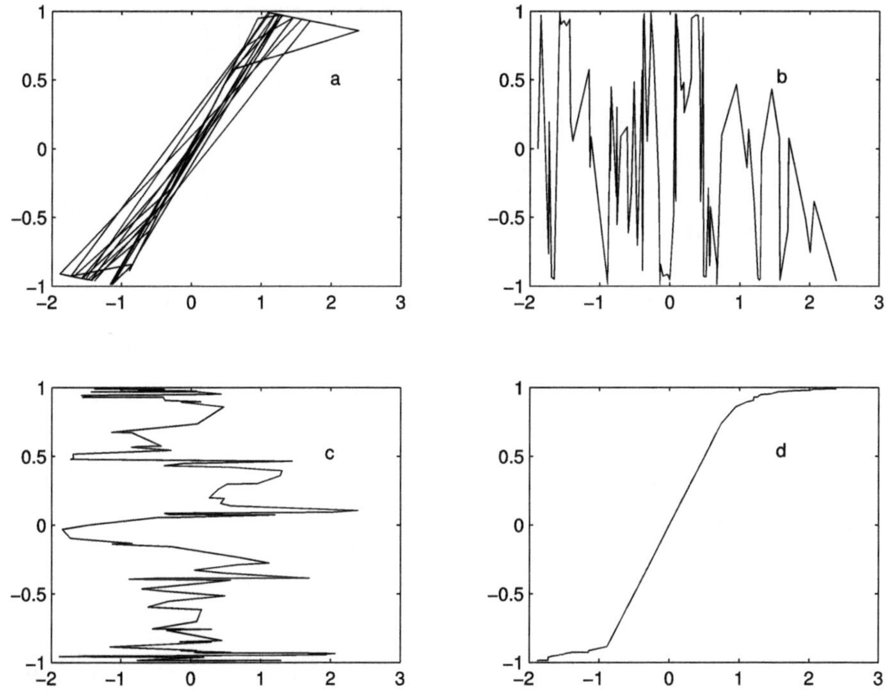

Fig. 7.17 The values, notation and markings are the same as in Fig. 7.7

some observed threshold variable is smaller or larger than d. Therefore the problem of the identification of unknown parameters of nonlinear and linear blocks of the Hammerstein systems could be solved, if a simple way of partitioning the available data sets were found in the case of unknown d. Afterwards, the estimates of parameters of regression functions could be calculated by processing particles of observations to be determined just like they were calculated in the previous section.

Let us rearrange the unnoisy open-loop Hammerstein system (Fig. 1.2) output data $y(k)\ \forall\ k \in \overline{1, \mathrm{N}}$ that, firstly, was processed by inverse transfer function of the linear block $G(q^{-1}, \mathbf{\Theta})$ as follows

$$\hat{x}(k) = G^{-1}(q^{-1}, \hat{\mathbf{\Theta}}_{\circ})y(k), \tag{7.41}$$

in an ascending order of their values. As a result, the Hammerstein system internal signal estimate $\hat{x}(k)$ is reconstructed in Eq. (7.41) if measurement noise $v(k)$ is absent. Here $\hat{\mathbf{\Theta}}_{\circ}$ is vector of initial values of unknown parameters. Note that the filtering of $y(k)$ is possible if the linear block of the Hammerstein system is invertible and parameter values of $G^{-1}(q^{-1}, \hat{\mathbf{\Theta}}_{\circ})$ are known, even approximately.

Thus, three Hammerstein system internal signal versions could be available in the mind: the initial intermediate signal $x(k)$; the preprocessed signal $\hat{x}(k)$ in Eq. (7.41), and its reordered version $\tilde{x}(k)$. Assuming that the threshold of PWL nonlinearity is

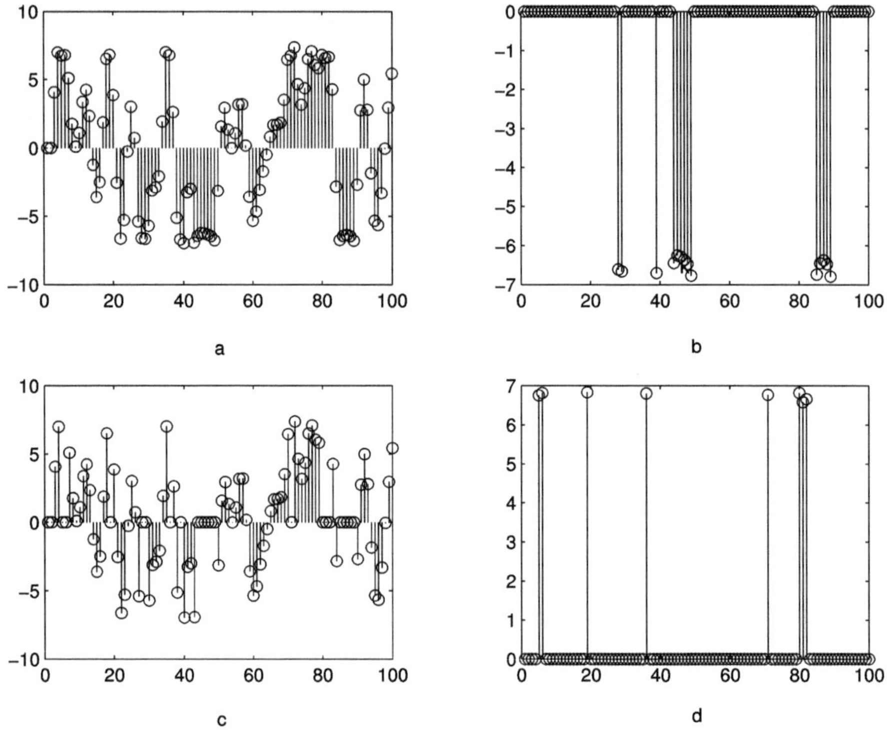

Fig. 7.18 Samples of $y(k)$ (**a**) (see Fig. 7.24c) and its data sets: left (**b**), middle (**c**), right (**d**)(here the observations, that belong to the other data set, are equal zeros). Input $u(k)$ of the form (7.31) (see Fig. 7.14a)

known, the observation samples of the rearranged intermediate signal $\tilde{x}(k)$ should be partitioned into three data sets: left-hand side data set (N_1 samples) with values lower than or equal to negative d value, middle data set (N_3 samples) with values higher than negative threshold value d but lower or equal to positive d value, and right-hand side data set (N_2 samples) with values higher than threshold d. So, $N = N_1 + N_3 + N_2$ is the sample size of the intial sequence $y(k)$.

If the threshold is unknown then the dependence of sort($\hat{y}(k)$) on the input sort($u(k)$) can help us (Fig. 1.2) to determine the critical poins of sequence $\tilde{y}(k)$.

From the engineering point of view it is assumed that no less than 80% observations of the sequence $\tilde{y}(k)$ are concentrated on the middle-set and approximately by 10% or less on any side set. Hence, the observations of the rearranged input $\tilde{u}(k)$ with the highest and positive values will be concentrated on the right-hand side set, while the observations with the lowest and negative values on the left-hand side one. Thus, the middle data set $\tilde{u}(k)\ \forall\ k \in \overline{\hat{N}_1 + l_1, \hat{N}_3 - l_2}$ is, really, reordered in an ascending order of their values $u(k)$ having, really, some portions of missing observations within it that belong to the left-hand and right-hand side sets of the data.

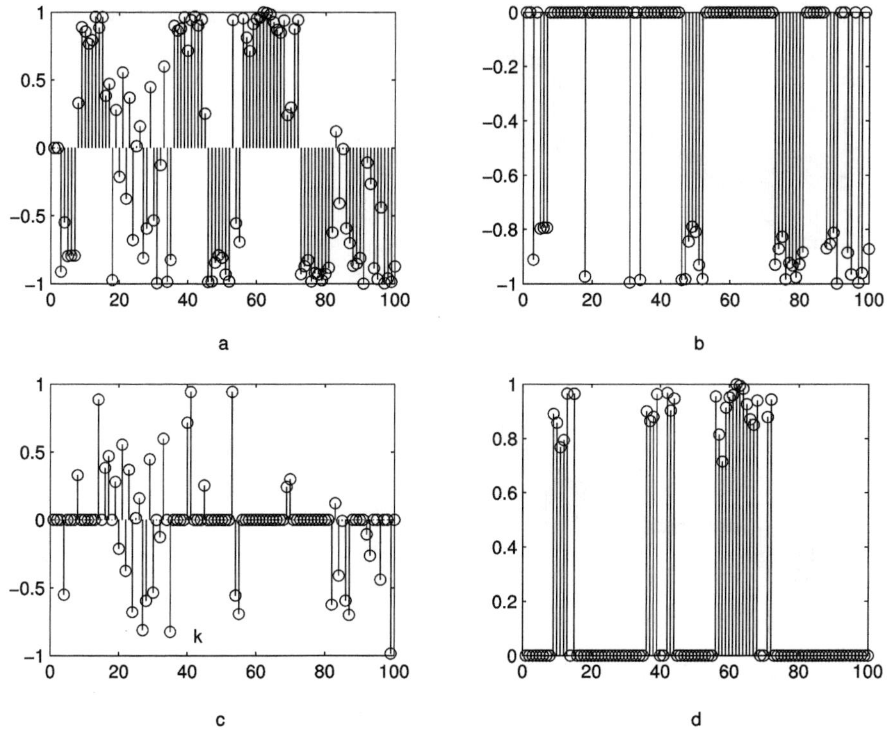

Fig. 7.19 Samples of signal $y(k)$ (**a**) (see Fig. 7.15c) and its data sets: left (**b**), middle (**c**), right (**d**). Input $u(k)$ is white Gaussian noise (see Fig. 7.15a)

Let us suppose now that the process noise $w(k)$ is absent (Fig. 1.2). Then the observations of the unknown intermediate signal $x(k)$ are coincidental with the respective observations of the middle data set $\tilde{u}(k)$, equivalent to those input observations $u(k)$ that passed the PWL nonlinearity (7.6) without any processing, avoiding parasitical distortions. In such a case, one could get these observations simply by choosing the upper interval bound slightly lower than the 80% and the lower interval bound negligibly higher than the 10% of the sampled reordered observations of $\tilde{u}(k)$.

Let us now partially reconstruct an unmeasurable intermediate signal $x(k)$, choosing in the initial order only those values of $u(k)$ $\forall\, k \in \overline{1, \mathrm{N}}$ that are present in the middle data set of $\tilde{u}(k)$ $\forall\, k \in \overline{\mathrm{N}_1 + l_1, \mathrm{N}_3 - l_2}$. Here l_1, l_2 are some portions of observations of the middle data set of $\tilde{u}(k)$. In such a case one could get $x(k) \equiv u(k)$ $\forall\, k \in \overline{\mathrm{N}_1 + l_1, \mathrm{N}_3 - l_2}$. Really, if the process noise $v(k)$ is absent, then the intermediate sequence $x(k)$ is equivalent sequence $\tilde{y}(k)$ with some small portions of observations in it that belong to the left-hand or right-hand side sets of the rearranged data. It could be used to calculate the estimates of parameters (7.3) of the transfer function $G(q^{-1}, \boldsymbol{\Theta})$ according to

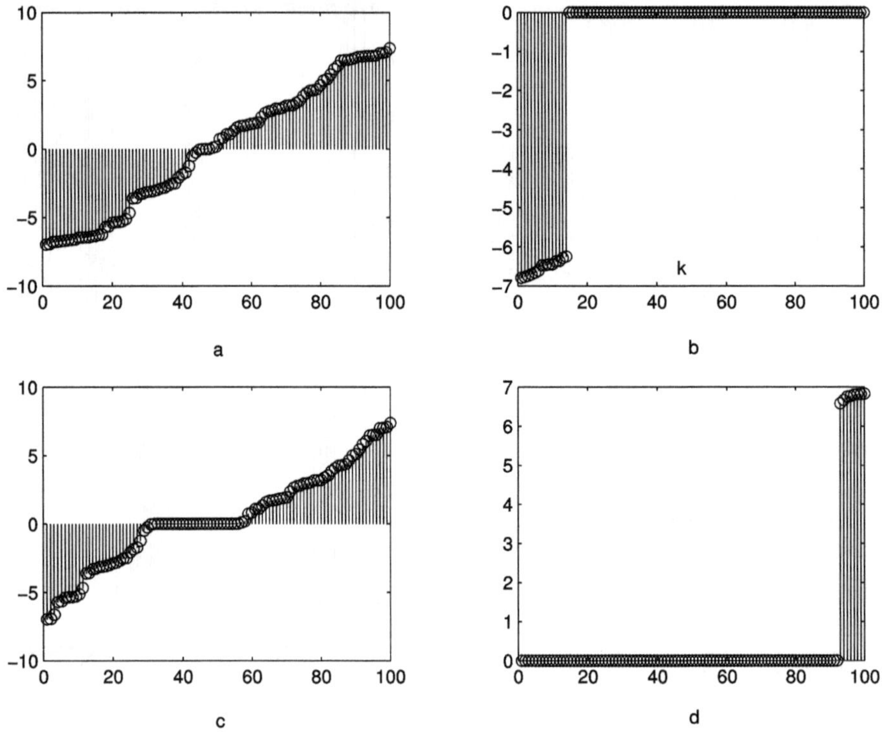

Fig. 7.20 The reordered in an ascending order of their values signal $y(k)$ (**a**) (see Figs. 7.14c and 7.18a) and its rearranged data sets: left (**b**), middle (**c**), right (**d**) (here the observations, that belong to the other data set, are equal to zeros). Input $u(k)$ of the form (7.31) (see Fig. 7.14a)

$$\hat{\mathbf{\Theta}} = \left(\tilde{\mathbf{\Psi}}^T \tilde{\mathbf{\Psi}}\right)^{-1} \tilde{\mathbf{\Psi}}^T \tilde{\mathbf{Y}}, \tag{7.42}$$

when the initial system of linear equations is

$$\tilde{\mathbf{Y}} = \tilde{\mathbf{\Psi}} \mathbf{\Theta}, \tag{7.43}$$

where

$$\tilde{\mathbf{\Psi}} = \begin{bmatrix} \tilde{u}(0) & \ldots \tilde{u}(1-n) & \hat{x}(0) & \ldots \hat{x}(1-n) \\ \tilde{u}(1) & \ldots \tilde{u}(2-n) & \hat{x}(1) & \ldots \hat{x}(2-n) \\ \vdots & \vdots & \vdots & \vdots \\ \tilde{u}(k-1) & \ldots \tilde{u}(k-n) & \hat{x}(k-1) & \ldots \hat{x}(k-n) \end{bmatrix} \tag{7.44}$$

is the $k \times 2n$ matrix, consisting of observation samples of the reordered input $\tilde{u}(k)$ and reordered internal signal $\tilde{y}(k)$ values of which are associated with $\tilde{u}(k)$; $\tilde{\mathbf{Y}} = (\tilde{y}(1), \tilde{y}(2), \ldots, \tilde{y}(k))^T$ is the $k \times 1$ vector, consisting of the samples of reordered output $y(k)$ values that correspond to the middle side set of the reordered data.

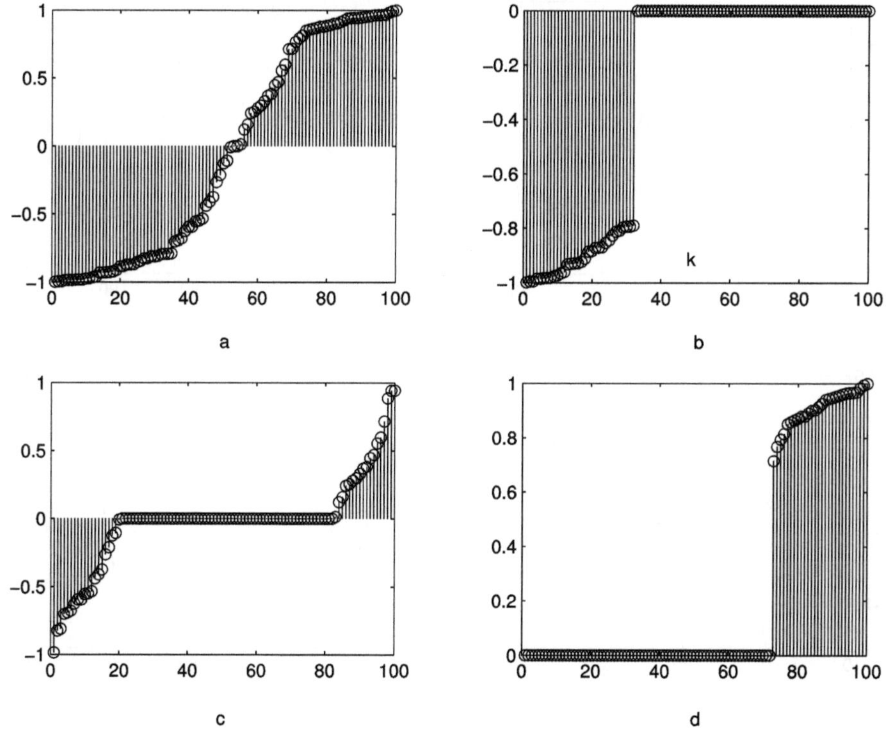

Fig. 7.21 The reordered in an ascending order of their values signal $y(k)$ (**a**) (see Figs. 7.15c and 7.27a) and its rearranged data sets: left (**b**), middle (**c**), right (**d**). Input $u(k)$ is white Gaussian noise (see Fig. 7.15a)

Here

$$\hat{\boldsymbol{\Theta}}^T = \left(\hat{\mathbf{b}}, \hat{\mathbf{a}}\right)^T, \quad \hat{\mathbf{b}}^T = \left(\hat{b}_1, \ldots, \hat{b}_n\right), \quad \hat{\mathbf{a}}^T = \left(\hat{a}_1, \ldots, \hat{a}_n\right) \tag{7.45}$$

are $2n \times 1$, $n \times 1$, $n \times 1$ vectors of parameter estimates (7.3), respectively.

Estimates of the parameters c_0, d_0 and c_1, d_1 are calculated by the ordinary LS, too. In such a case, criteria:

$$I(c_0, c_1) = \frac{1}{2\mathrm{N}_1} \sum_{i=1}^{\mathrm{N}_1} \left[\hat{x}(i) - c_0 - c_1 \tilde{u}(i)\right]^2, \tag{7.46}$$

$$I(d_0, d_1) = \frac{1}{2\mathrm{N}_2} \sum_{j=1}^{\mathrm{N}_2} \left[\hat{x}(j) - d_0 - d_1 \tilde{u}(j)\right]^2, \tag{7.47}$$

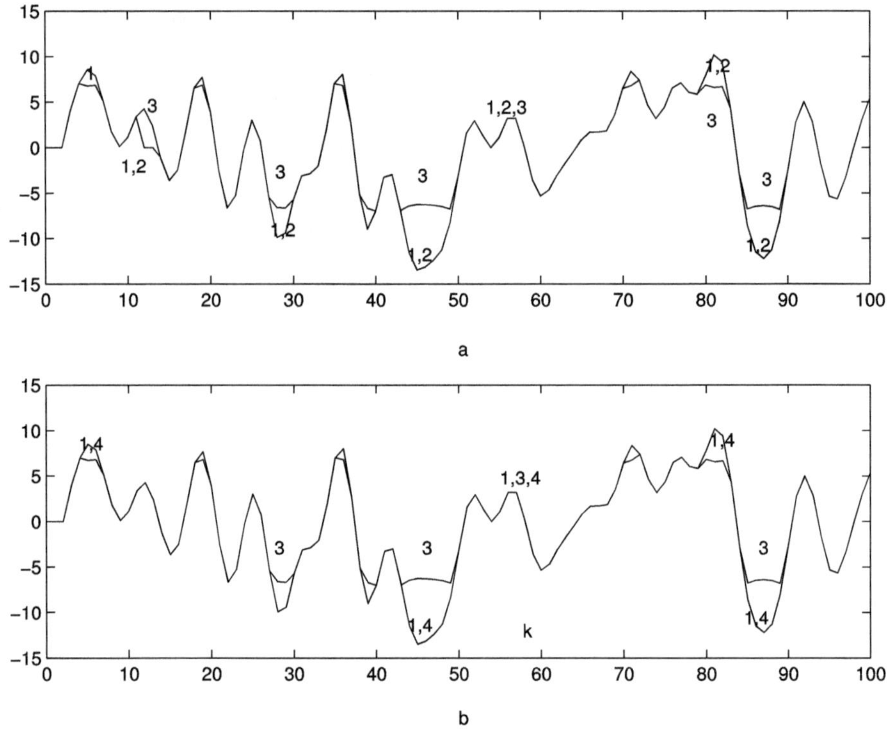

Fig. 7.22 The intermediate signal $x(k)$ (curve 1), the output signal $y(k)$ (curve 3), the reconstructed versions of $x(k)$ (curves 2, 4), calculated using Eq. (7.21) (**a**) and Eq. (7.34) (**b**), respectively. Input $u(k)$ of the form (7.31) (see Fig. 7.14a)

will be minimized with respect to parameters c_0, c_1 and d_0, d_1, correspondingly, using left- and right-hand side sets data particles of $\hat{x}(k)$ and respective observations of the rearranged input signal $\tilde{u}(k)$. Here $\hat{x}(k)$ are observations of the intermediate signal that were rearranged in accordance with $\tilde{u}(k)$.

The estimates of parameters c_1, d_1 and c_0, d_0 are calculated according to [8]

$$\hat{c}_1 = \frac{\sum_{j=1}^{N_1} \hat{x}(i)\tilde{u}(i)}{\sum_{i=1}^{N_1} \tilde{u}^2(i)}, \qquad \hat{d}_1 = \frac{\sum_{j=1}^{N_2} \hat{x}(j)\tilde{u}(j)}{\sum_{j=1}^{N_2} \tilde{u}^2(j)}, \tag{7.48}$$

$$\hat{c}_0 = \frac{\sum_{i=1}^{N_1} [\hat{x}(i) - \hat{c}_1\tilde{u}(i)]}{N_1}, \qquad \hat{d}_0 = \frac{\sum_{j=1}^{N_2} [\hat{x}(j) - \hat{d}_1\tilde{u}(j)]}{N_2}, \tag{7.49}$$

respectively, but using side-sets data particles of $\tilde{u}(k)$ and associated observations of the intermediate signal $\hat{x}(k)$, that are reordered in accordance with $\tilde{u}(k)$.

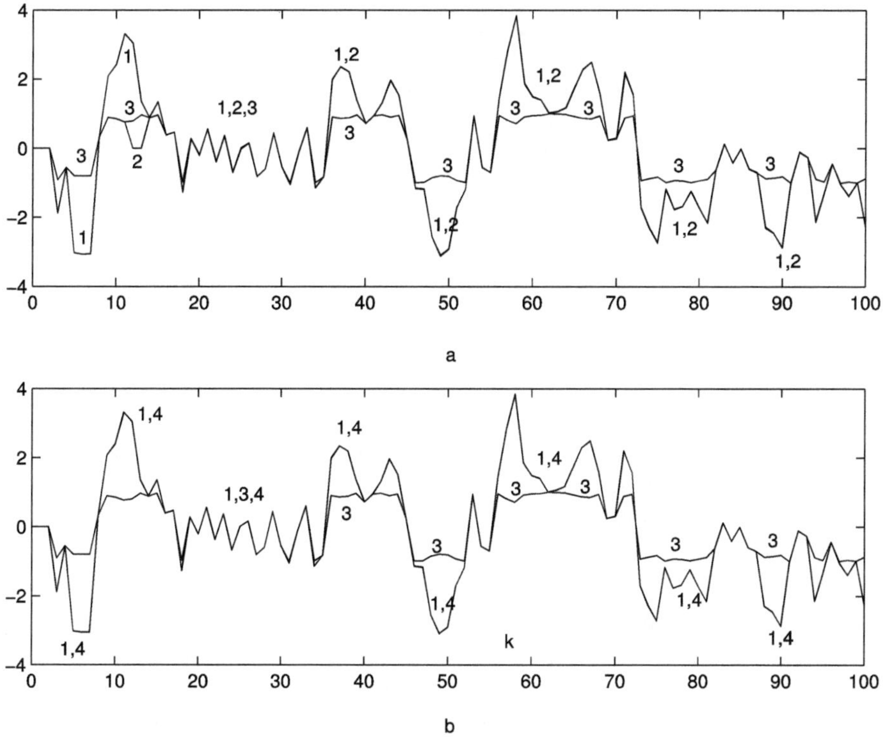

Fig. 7.23 The intermediate signal $x(k)$ (curve 1), the output signal $y(k)$ (curve 3), the reconstructed versions of $x(k)$ (curves 2, 4), calculated using Eq. (7.21) (**a**) and Eq. (7.34) (**b**), respectively. Input $u(k)$ is white Gaussian noise (see Fig. 7.15a)

7.6 Numerical Simulation for the Hammerstein System

The true non-noisy intermediate signal $x(k)$ $k = \overline{1, \mathrm{N}}$, of the Hammerstein system (Figs. 7.24b and 7.25b) is given according to Pupeikis [12] by

$$x(k) = \begin{cases} -0.9 + 0.1u(k) & \text{if } u(k) \leq -1, \\ u(k) & \text{if } -1 < u(k) \leq 1, \\ 0.9 + 0.1u(k) & \text{if } u(k) > 1 \end{cases} \tag{7.50}$$

with the sum of sinusoids (Fig. 7.24a)

$$u(k) = \frac{1}{20} \sum_{i=1}^{20} \sin(i\pi k/10 + \phi_i) \tag{7.51}$$

Table 7.5 Averaged estimates of the parameters $b_1, a_1, c_0, c_1, d_0, d_1$, and thresholds a, $-a$ with their confidence intervals. Input: the periodical signal (7.31). SNR^v=10

Estimates	SNR^e = 1	SNR^e = 10	SNR^e = 100
$\hat{b}_1$	1.42 ± 0.33	1.15 ± 0.03	1.13 ± 0.00
$\hat{a}_1$	-0.68 ± 0.05	-0.68 ± 0.01	-0.68 ± 0.00
$\hat{c}_0$	-7.71 ± 0.07	-7.86 ± 0.65	-8.47 ± 4.9
$\hat{c}_1$	0.02 ± 0.012	0.01 ± 0.11	0.27 ± 0.72
$\hat{d}_0$	6.49 ± 0.11	6.49 ± 1.14	3.11 ± 9.28
$\hat{d}_1$	0.09 ± 0.01	0.07 ± 0.05	-0.03 ± 0.39
$\hat{a}$	6.62 ± 0.04	6.5 ± 0.45	13.71 ± 12.51
$-\hat{a}$	-7.09 ± 0.1	-7.42 ± 1.11	1.81 ± 24.4

Table 7.6 The values and notation are the same as in Table 7.3. Input—the Gaussian white noise

Estimates	SNR^e = 1	SNR^e = 10	SNR^e = 100
$\hat{b}_1$	0.79 ± 0.21	0.95 ± 0.02	0.97 ± 0.00
$\hat{a}_1$	-0.48 ± 0.17	-0.71 ± 0.01	-0.72 ± 0.00
$\hat{c}_0$	-1.03 ± 0.00	-1.04 ± 0.03	-1.17 ± 0.17
$\hat{c}_1$	0.12 ± 0.00	0.12 ± 0.02	0.24 ± 0.3
$\hat{d}_0$	1.09 ± 0.01	1.07 ± 0.05	0.89 ± 0.4
$\hat{d}_1$	-0.1 ± 0.00	-0.09 ± 0.03	-0.02 ± 0.32
$\hat{a}$	0.91 ± 0.00	0.92 ± 0.04	1.02 ± 0.28
$-\hat{a}$	-0.99 ± 0.00	-0.98 ± 0.02	0.81 ± 0.29

and white Gaussian noise with variance 1.7 (Fig. 7.25a) as input or training signals to the nonlinear block. In (7.51) the stochastic variables ϕ_k with a uniform distribution on $[0, 2\pi]$ were chosen. The true output signal (Figs. 7.24d and 7.25d) is described by

$$y(k) = G(q^{-1}, \mathbf{\Theta})x(k) = \frac{b_1 q^{-1}}{1 + a_1 q^{-1}} x(k). \tag{7.52}$$

with true values $b_1 = 0.3$ and $a_1 = -0.5$ while initial values are $b_1 = 0.1$ and $a_1 = -0.25$. First of all, N = 100 data points have been generated without additive process and measurement noises (Figs. 7.24 and 7.25). Afterwards, the LS problem (7.42) was solved, using 56 and 55 rearranged observations of the input, respectively (Figs. 7.26c and 7.27c), excluding zeros, and associated observations of the respective outputs (Figs. 7.24d and 7.25d). The estimate $\hat{x}(k)$ of the intermediate signal $x(k)$ was reconstructed according to

$$\hat{x}(k) = \frac{1}{\hat{b}_1} y(k+1) + \frac{\hat{a}_1}{\hat{b}_1} y(k), \tag{7.53}$$

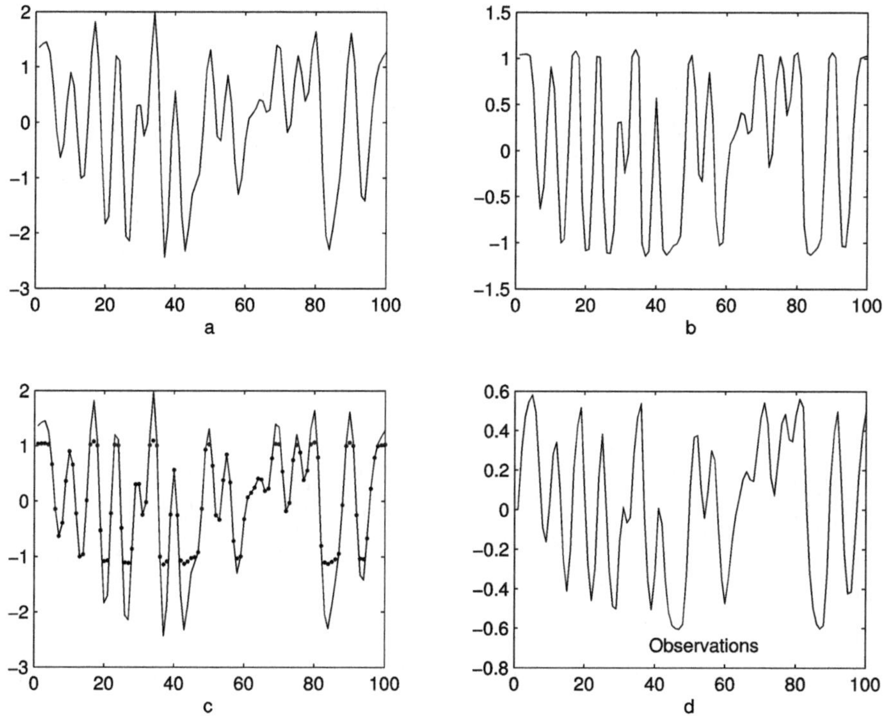

Fig. 7.24 The signals of the simulated Hammerstein system with a piecewise nonlinearity (7.50): input $u(k)$, calculated by (7.51) (**a**), intermediate signal $x(k)$ (**b**), input $u(k)$ and intermediate signal $x(k)$ (dotted line) (**c**) and output $y(k)$ (**d**)

replacing unknown true values of parameters by their estimates. In such a case, the estimates $\hat{b}_1, \hat{a}_1$ were approximately equal to the true parameters: $b_1 = 0.3, a_1 = 0.5$. The reconstructed versions of the intermediate signal $x(k)$ are shown in (Figs. 7.30a, b). It ought to be noted that the accuracy of estimates of the intermediate signal, calculated by formula (7.53), for both inputs is the same, when the process- and measurement noises are absent. If $\hat{x}(k)$ has been obtained, then it is simple to separate different particles of observations of $\hat{x}(k)$ that belong to the respective side-sets of reordered $u(k)$ (Figs. 7.26, 7.27, 7.28 and 7.29). The estimates of parameters c_1, d_1 and c_0, d_0 are calculated according to formulas (7.48) and (7.49), respectively. In such a case, the rearranged observations of $\hat{x}(k)$ and $u(k)$ were substituted in formulas (7.48) and the estimates of c_1 and d_1 were determined: $\hat{c}_1 = \hat{d}_1 = 0.1$. Then, the estimates $\hat{c}_0$ and $\hat{d}_0$ were calculated by (7.49). Their values are also coincidental with the values of true coefficients: $\hat{c}_0 = -0.9$, while $\hat{d}_0 = 0.9$. It should be noted that $N_1 = 19$, $N_2 = 23$ for the periodical signal (7.51) (Fig. 7.24a) and $N_1 = 25$, $N_2 = 20$ for the Gaussian white noise (Fig. 7.25a) were used to calculate the estimates $\hat{c}_0, \hat{c}_1, \hat{d}_0, \hat{d}_1$, Eqs. (7.48) and (7.49) to calculate the estimates $\hat{c}_0, \hat{c}_1, \hat{d}_0, \hat{d}_1$ were

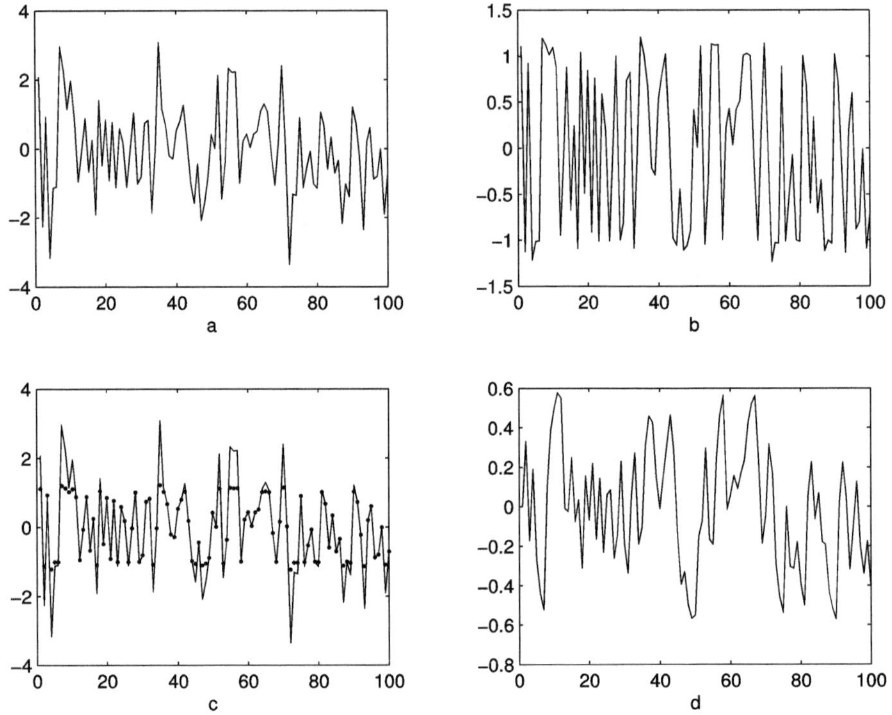

Fig. 7.25 The signals of the simulated Hammerstein system with a piecewise nonlinearity (7.50): input $u(k)$ is white Gaussian noise (**a**), intermediate signal $x(k)$ (**b**), input $u(k)$ and intermediate signal $x(k)$ (dotted line) (**c**) and output $y(k)$ (**d**)

used. The values of estimates $\hat{a}$ were equal to the true value $a = 1$. Note that the recursive approach is possible here according to *V*ö**rö**s [19], too.

In order to determine how realizations of different process- and measurement noises affect the accuracy of estimation of unknown parameters, we have used the Monte Carlo simulation with 10 data samples, each containing 100 pairs of input-output observations. 10 experiments with the same realization of the process noise $v(k)$ and different realizations of the measurement noise $e(k)$ with different levels of its intensity have been carried out. The intensity of noises was assured by choosing respective SNR (the square root of the ratio of signal and noise variances). For the process noise, SNR^{v} was equal to 100, and for the measurement noise, SNR^{e}: 1, 10, 100. As inputs for all given nonlinearities the periodical signal (7.51) and white Gaussian noise were chosen. In each ith experiment the estimates of parameters were calculated. During the Monte Carlo simulation averaged values of estimates of the parameters and of the threshold and their confidence intervals were calculated. In Tables 7.7 and 7.8, for each input the averaged estimates of parameters and the

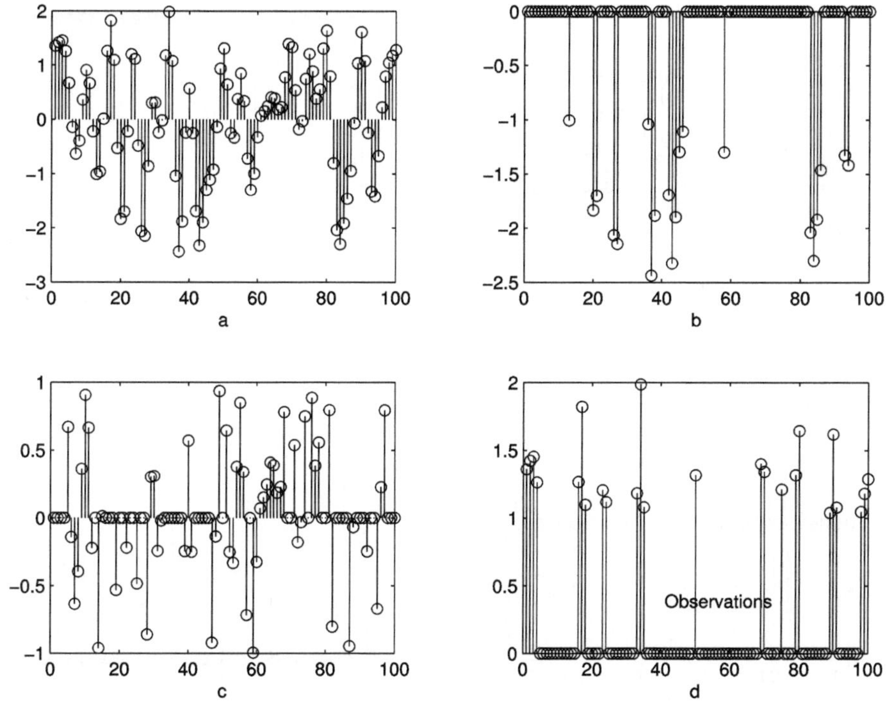

Fig. 7.26 Samples of $u(k)$ (**a**) (see Fig. 7.24a) and its data sets: left (**b**), middle (**c**), right (**d**) (here the observations, that belong to the other data set, are equal zeros). Input $u(k)$ of the form (7.51)

threshold a of the simulated Hammerstein system (Fig. 1.2) with the linear part (7.52) ($b_1 = 0.3$; $a_1 = -0.5$) and the PWL nonlinearity (7.50) ($c_0 = -0.9$, $c_1 = 0.1$, $d_0 = 0.9$, $d_1 = 0.1$) with their confidence intervals are presented. It ought to be noted that in each experiment here the value of SNR^v was fixed and was the same, while the values of SNR^e were varying due to different realizations of $e(k)$. The Monte Carlo simulation (Tables 7.7 and 7.8) implies that the accuracy of parametric identification of the Hammerstein system depends on the intensity of measurement noise.

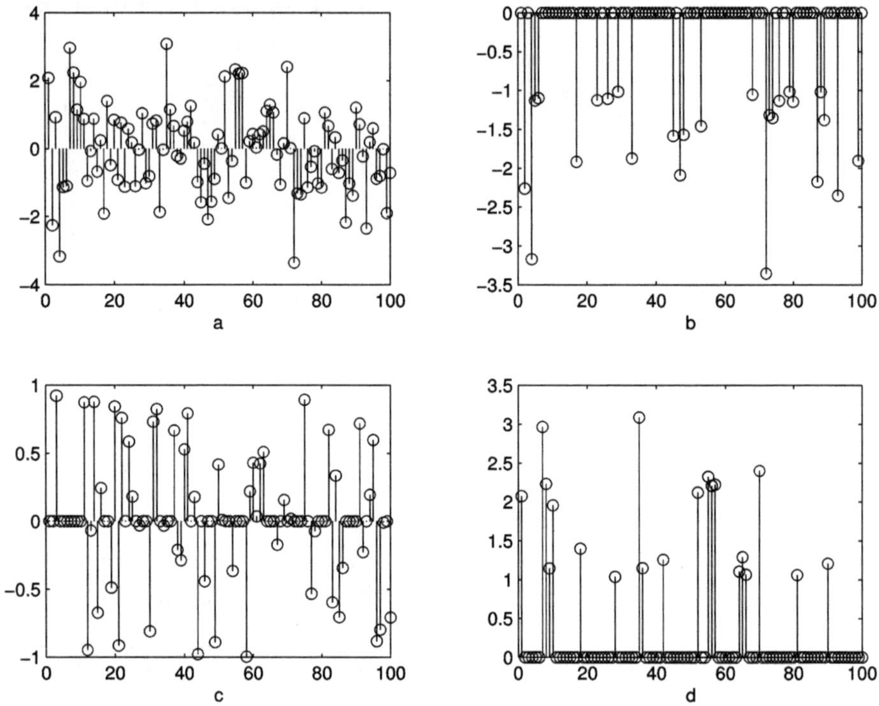

Fig. 7.27 Samples of signal $u(k)$ (**a**) (see Fig. 7.25a) and its data sets: left (**b**), middle (**c**), right (**d**). Input $u(k)$ is white Gaussian noise

7.7 Conclusions

It is shown here that a problem of parametrical identification of block-oriented switching systems could be essentially reduced by a simple data rearrangement in an ascending order according to their values. Thus, the available data are partitioned into the known number of data sets, respectively the different number of system functioning regimes, that correspond to the distinct threshold regression models. Afterwards, the unknown parameter estimates of linear regression models can be calculated by processing respective sets of the rearranged output and associated input observations. A technique, based on ordinary LS, is proposed here for estimating the parameters of linear and nonlinear parts of the Wiener and Hammerstein systems, including the unknown threshold of the PWL nonlinearity, too. During successive steps the unknown intermediate signal is reconstructed and the missing values of observations of respective data particles are replaced by their estimates. Various results of numerical simulation (Figs. 7.4–7.30), including that of Monte Carlo (Tables 7.5–7.7 and 7.8) are obtained. Parameter and threshold estimates, determined for the Wiener and Hammerstein systems having identical linear and nonlinear parts, correspond to the respective true parameter and threshold values, if the assumption that no less than

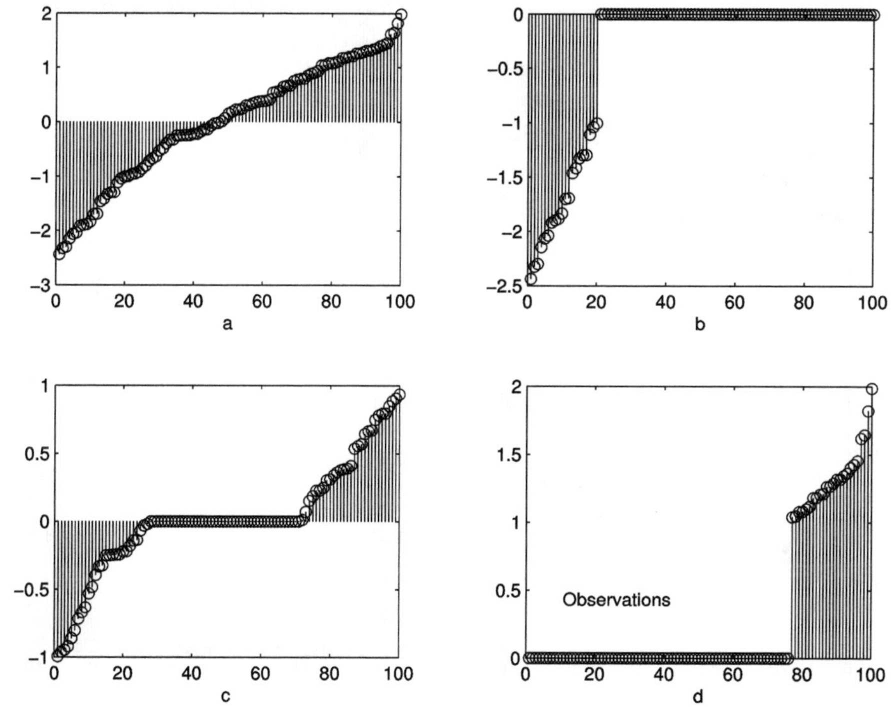

Fig. 7.28 The reordered in an ascending order of their values signal $u(k)$ (**a**) (see Fig. 7.26a) and its rearranged data sets: left (**b**), middle (**c**), right (**d**) (here the observations, that belong to the other data set, are equal zeros)

80% observations are concentrated on the middle-set and approximately by 10% or less on any side set is valid. On the other hand, in the noisy frame the artificial correlating of process noise $w(k)$ by the linear part of the Hammerstein system or that by the PWL nonlinearity of the Wiener system have an effect upon Wiener and Hammerstein systems parameter estimates. In such a case, the estimates, obtained for the Wiener system were more accurate, than estimates derived for the Hammerstein one, under the same circumstances.

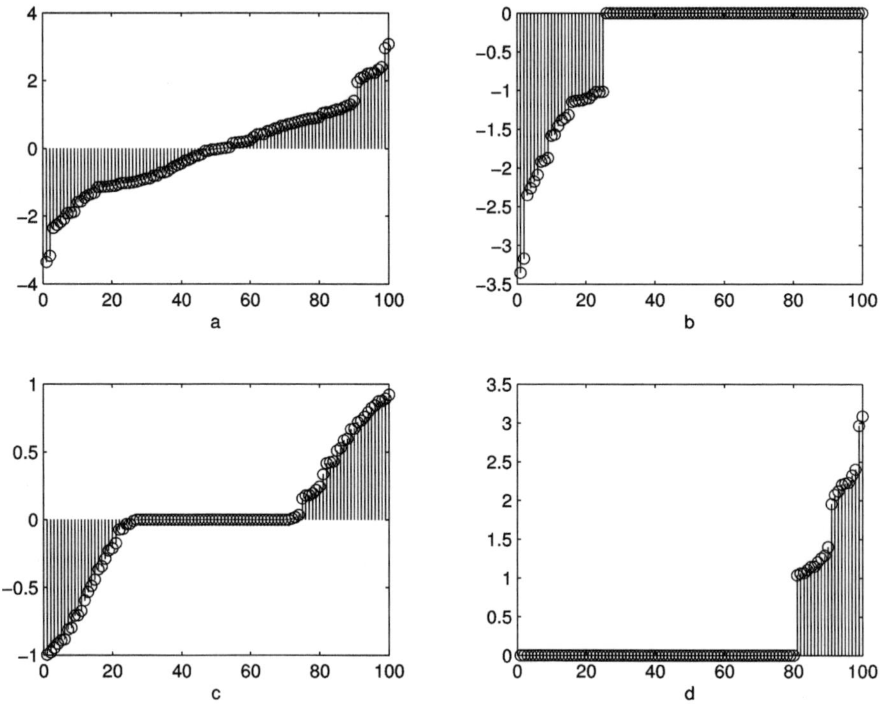

Fig. 7.29 The reordered in an ascending order of their values signal $u(k)$ (**a**) (see Fig. 7.27a) and its rearranged data sets: left (**b**), middle (**c**), right (**d**)

Table 7.7 Averaged estimates of the parameters $b_1, a_1, c_0, c_1, d_0, d_1$, and thresholds a, $-a$ with their confidence intervals. Input: the periodical signal. $\mathrm{SNR}^v=100$

Estimates	$\mathrm{SNR}^e = 1$	$\mathrm{SNR}^e = 10$	$\mathrm{SNR}^e = 100$
$\hat{b}_1$	0.39 ± 0.10	0.32 ± 0.03	0.3 ± 0.01
$\hat{a}_1$	-0.09 ± 0.09	-0.39 ± 0.04	-0.48 ± 0.01
$\hat{c}_0$	-0.99 ± 0.03	-1.04 ± 0.09	-1.20 ± 0.39
$\hat{c}_1$	0.06 ± 0.02	0.04 ± 0.07	0.02 ± 0.24
$\hat{d}_0$	0.78 ± 0.1	0.55 ± 0.30	0.19 ± 0.83
$\hat{d}_1$	0.15 ± 0.07	0.31 ± 0.21	0.57 ± 0.61
$\hat{a}$	0.91 ± 0.06	0.57 ± 0.55	0.62 ± 0.83
$-\hat{a}$	-1.05 ± 0.02	-1.09 ± 0.08	-1.23 ± 0.36

Table 7.8 The values and notation are the same as in Table 7.7. Input—the Gaussian white noise

Estimates	$\text{SNR}^e = 1$	$\text{SNR}^e = 10$	$\text{SNR}^e = 100$
$\hat{b}_1$	0.34 ± 0.05	0.32 ± 0.01	0.3 ± 0.00
$\hat{a}_1$	-0.17 ± 0.11	-0.42 ± 0.04	-0.48 ± 0.02
$\hat{c}_0$	-0.93 ± 0.05	-0.95 ± 0.16	-1.08 ± 0.51
$\hat{c}_1$	0.07 ± 0.03	0.07 ± 0.08	0.02 ± 0.28
$\hat{d}_0$	0.89 ± 0.04	0.91 ± 0.13	0.94 ± 0.34
$\hat{d}_1$	0.09 ± 0.02	0.08 ± 0.07	0.07 ± 0.18
$\hat{a}$	0.98 ± 0.03	0.98 ± 0.08	0.97 ± 0.22
$-\hat{a}$	-1 ± 0.03	-1 ± 0.09	-1.05 ± 0.25

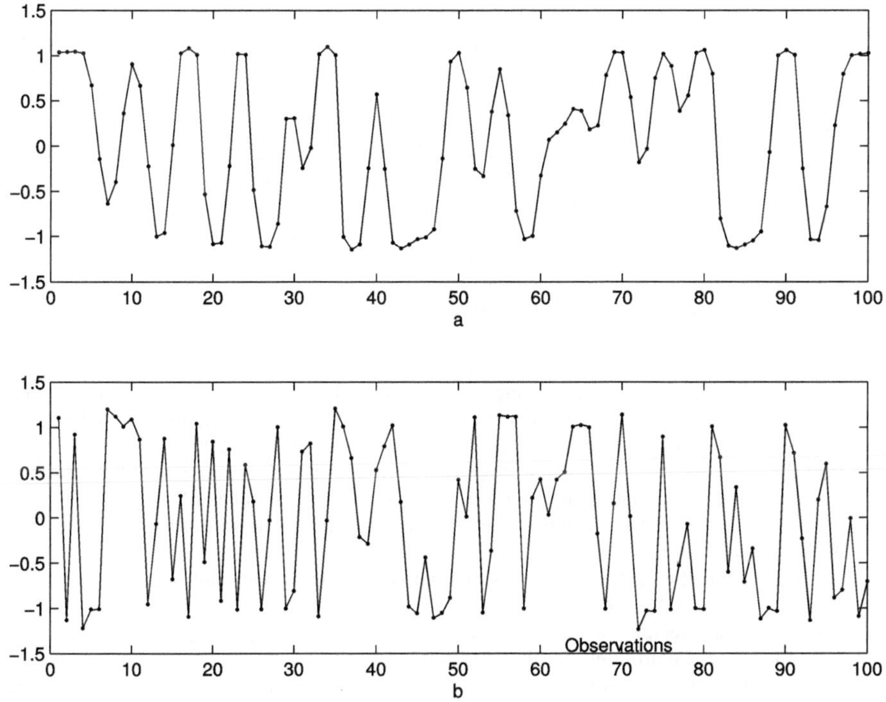

Fig. 7.30 The intermediate signal $x(k)$ (continuous line), and the reconstructed versions of $x(k)$ (dotted line), calculated using Eq. (7.53) with the sum of sinusoids (Fig. 7.24a) (**a**) and white Gaussian noise (Fig. 7.25a) (**b**)

References

1. *Å*str*ö*m, K.J., Wittenmark, B.: Computer Controlled Systems. Prentice-Hall Inc, Englewood Cliffs (1984)
2. Bloemen, H.H.J., Chou, C.T., Van den Boom, T.J.J., Verdult, V., Verhaegen, M., Backx, T.C.: Wiener model identification and predictive control for dual composition control of a distillation column. J. Proc. Control **11**, 601–620 (2001)
3. Glad, T., Ljung, L.: Control Theory, Multivariable and Nonlinear Methods. Taylor & Francis (2000)
4. Hagenblad, A.: Aspects of the Identification of Wiener Models. Licentiate thesis No. 793. Department of Electrical Engineering, Link*öping, Sweden, Nov 1999* (1999)
5. Hansen, B.E.: Threshold effects in non-dynamic panels: estimation, testing, and inference. J. Econometrics **93**(1999), 345–368 (1999)
6. Lankaster, P., Tsimenetsky, M.: The Theory of Matrices: With Applications. Academic, San Diego (1985)
7. Ljung, L.: Convergence analysis of parametric identification methods. IEEE Trans. AC **AC-23**, 770–783 (1978)
8. Malinvaud, E.: Methodes statistiques de L'Econometrie. , Paris (1969)
9. Eykhoff, P.: System Identification, Parameter and State Estimation. Wiley, NJ (1974)
10. Pupeikis, R.: Identification of piecewise affine Wiener systems using data partition. Technical Report LiTH-ISY-R-2523, Department of Electrical Engineering, Linköping University, SE-581 83 Linköping, Sweden (2003)
11. Pupeikis, R.: On the identification of Wiener systems having saturation-like functions with positive slopes. Informatica, LT MII VU Press Vilnius **16**(1), 131–144 (2005)
12. Pupeikis, R.: On the identification of Hammerstein systems having saturation-like functions with positive slopes. Informatica, LT MII VU Press Vilnius **17**(1), 55–68 (2006)
13. Pupeikis, R.: Parametric identification of linear system followed by non-invertible piecewise nonlinearities. Int. Front. Sci. Lett. **6**(1), 16–27 (2015)
14. Pupeikis, R., Navakauskas, D., Ljung, L.: Identification of Wiener systems with hard and discontinuous nonlinearities. Technical Report LiTH-ISY-R-2501, Department of Electrical Engineering, Linköping University, SE-581 83 Linköping, Sweden (2003)
15. Roll, J.: Local and Piecewise Affine Approach to System Identification, Ph.D. thesis No 802. Department of Electrical Engineering, Linköping University, Sweden (2003)
16. Tao, G., Kokotovi*č*, P.V.: Adaptive control of plants with unknown dead-zones. Trans. AC **39**(1), 59–68 (1994)
17. V*ö*r*ö*s, J.: Modeling and parameter identification of systems with multisegment piecewise-linear onlinearities. Syst. Control Lett. **47**, 184–188 (2002)
18. V*ö*r*ö*s, J.: Parameter identification of Wiener systems with multisegment piecewise-linear nonlinearities. Syst. Control Lett. **56**, 99–105 (2007)
19. V*ö*r*ö*s, J.: Recursive identification of systems with noninvertible output nonlinearities. Informatica **21**(1), 139–148 (2010)
20. V*ö*r*ö*s, J.: Compound operator decomposition and its application to Hammerstein and Wiener systems. In: Giri, S.F., Bai E.-W. (eds.) Block-Oriented Nonlinear System Identification. Chapter 4, pp. 35–51 (2010b)
21. in Internet: online/stat.psu.edu./stat501/lesson/8/8.10/2023. Accessed in Dec 2023

Chapter 8
Parametric Identification of Systems with Deadzones

The approach, based on the observation samples reordering, is used here for the pre-processing the input-output data of the nonlinear block-oriented Wiener systems having the deadzone as PWL nonlinearity. Afterwards, it follows the parametrical identification of the linear block parameters. Numerical simulation results by computer are given.

8.1 Statement of the Problem

The Wiener system consists of a linear block, which takes precedence of a static nonlinearity (Fig. 1.3) (Hagenblad 1999); [6]. The relationship between an output $y(k)$ and an internal signal $x(k)$ is

$$y(k) = f(x(k), \eta) + v(k) \; \forall \; k = \overline{1, N}, \tag{8.1}$$

where the nonlinear part of the Wiener system $f(x(k), \eta)$ is a deadzone (Fig. 8.1):

$$f(x(k), \eta) = \begin{cases} x(k) - d & \text{if } x(k) \geq d, \\ 0 & \text{if } -d < x(k) < d, \\ x(k) + d & \text{if } x(k) \leq d. \end{cases} \tag{8.2}$$

Here d is a threshold of the nonlinearity; the process noise $w(k)$ is hidden in the internal signal $x(k) \; \forall \; k = \overline{1, N}$.

The objective consists of parametric identification of the linear part of Wiener system in the presence of a deadzone.

R. Pupeikis and K. Kazlauskas, *Advanced Digital Signal Processing Methods for Filtering, Identification, and Nonlinear Systems Control*, Studies in Systems, Decision and Control 638, https://doi.org/10.1007/978-3-032-06740-1_8

8.2 Reordering of Observation Samples

The parametrical identification method based on the data reordering (see Sect. 7.2) will be used here for a deadzone as well. The FIR model (7.8) will be applied here to calculate the auxiliary signal $\hat{x}(k)$, i.e. the estimate of unmeasurable $x(k)$ of the Wiener system, by processing observations of the input $u(k)$ and partly damaged output $y(k)$ [2].

Let us rewrite the stored data (input $u(k)$ and output $y(k)\ \forall\ k = \overline{1, \mathrm{M}}$) samples in a matrix form as follows:

$$\mathbf{Y} = \boldsymbol{\Pi}\boldsymbol{\phi} + \mathbf{e}, \tag{8.3}$$

Here

$$\mathbf{Y} = (y(m), y(m+1), \ldots, y(\mathrm{M}-1), y(\mathrm{M}))^T \tag{8.4}$$

is $(\mathrm{N}+1) \times 1$ vector, consisting of output observation samples; $\mathbf{e} = (e(\mathrm{m}), e(\mathrm{m}+1), \ldots, e(\mathrm{M}-1), e\,(\mathrm{M})))^T$ is $(\mathrm{N}+1) \times 1$ vector of residuals;

$$\boldsymbol{\Pi} = \begin{bmatrix} u(\mathrm{m}-1) & u(\mathrm{m}-2) & \ldots & u(\mathrm{m}-\nu+1) & u(\mathrm{m}-\nu) \\ u(\mathrm{m}) & u(\mathrm{m}-1) & \ldots & u(\mathrm{m}-\nu+2) & u(\mathrm{m}-\nu+1) \\ \vdots & \vdots & \vdots & \vdots & \vdots \\ u(\mathrm{M}-1) & u(\mathrm{M}-2) & \ldots & u(\mathrm{M}-\nu+1) & u(\mathrm{M}-\nu) \end{bmatrix} \tag{8.5}$$

is the full rank $(\mathrm{N}+1) \times \nu$ regression matrix, consisting only of observations of the input $u(k)$;

$$\phi^T = \left(\phi_1, \phi_2 \ldots, \phi_{(\nu-1)}, \phi_\nu\right) \tag{8.6}$$

is $\nu \times 1$ vector of unknown parameters of FIR model (note that $\phi \equiv \mathbf{h}$ if coefficient $h_\circ$ is equal zero in Eq. (7.8)); $\mathrm{M} = \mathrm{N} + \mathrm{m}$.

In the previous section it was emphasized that the reason for the use of the FIR model consists of two important factors. Firstly, the dependence of some regressors on the process output will be facilitated, and, secondly, the assumption of ordinary LS that the regressors depend only on the non-noisy input signal, will be satisfied [1].

Next, let us rearrange the output data in the vector $\mathbf{V}$ in an ascending order of their values, reordering herewith the associated rows of the matrix $\boldsymbol{\Pi}$ as well. We do that by interchanging equations in the initial system (8.3) of linear equations [3].

It was noted earlier that the interchange of linear equations in a matrix system (8.3) does not influence the accuracy of LS parameter estimates to be calculated. Then, vector $\mathbf{V}$ and matrix $\boldsymbol{\Pi}$ should be partitioned into three data sets: the left-hand data set $\tilde{\mathbf{V}}_1 = \tilde{\boldsymbol{\Pi}}_1\tilde{\phi}_1$, the middle data set $\tilde{\mathbf{V}}_2 = \tilde{\boldsymbol{\Pi}}_2\tilde{\phi}_2$ and the right-hand data set $\tilde{\mathbf{V}}_3 = \tilde{\boldsymbol{\Pi}}_3\tilde{\phi}_3$ according to the three regimes of the deadzone nonlinearity. Here $\tilde{\mathbf{V}}_1, \tilde{\mathbf{V}}_2, \tilde{\mathbf{V}}_3$ are $(N_1+1) \times 1$, $(N_2+1) \times 1$ and $(N_3+1) \times 1$ vectors, respectively, while $\tilde{\boldsymbol{\Pi}}_1$, $\tilde{\boldsymbol{\Pi}}_2$ and $\tilde{\boldsymbol{\Pi}}_3$ are $(N_1+1) \times 1$, $(N_2+1) \times 1$ and $(N_3+1) \times 1$ matrices, correspondingly. Here $N = N_1 + N_3 + N_2$, moreover $N_1 = N_3$ for simplicity. Thus, initial system (8.3) is reordered into a system

$$\tilde{\mathbf{V}} = \tilde{\mathbf{\Pi}}\tilde{\phi}; \tag{8.7}$$

by simply interchanging equations. In (8.7):

$$\tilde{\mathbf{V}} = \begin{bmatrix} \tilde{\mathbf{V}}_1 \\ \tilde{\mathbf{V}}_2 \\ \tilde{\mathbf{V}}_3 \end{bmatrix}; \tilde{\mathbf{\Pi}} = \begin{bmatrix} \tilde{\mathbf{\Pi}}_1 \\ \tilde{\mathbf{\Pi}}_2 \\ \tilde{\mathbf{\Pi}}_3 \end{bmatrix}; \tilde{\phi} = \begin{bmatrix} \tilde{\phi}_1 \\ \tilde{\phi}_2 \\ \tilde{\phi}_3 \end{bmatrix}, \tag{8.8}$$

$$\tilde{\phi}_1^T = \left(\tilde{\phi}_{1,1}, \tilde{\phi}_{2,1} \ldots, \tilde{\phi}_{(\upsilon-1),1}, \tilde{\phi}_{\upsilon,1}\right), \tag{8.9}$$

$$\tilde{\phi}_2^T = \left(\tilde{\phi}_{1,2}, \tilde{\phi}_{2,2} \ldots, \tilde{\phi}_{(\upsilon-1),2}, \tilde{\phi}_{\upsilon,2}\right), \tag{8.10}$$

$$\tilde{\phi}_3^T = \left(\tilde{\phi}_{1,3}, \tilde{\phi}_{2,3} \ldots, \tilde{\phi}_{(\upsilon-1),3}, \tilde{\phi}_{\upsilon,3}\right), \tag{8.11}$$

Then, the left-hand side data set $\tilde{\mathbf{V}}_1$ (N_1 samples) consists of the reordered values $\tilde{y}(\tilde{k})\ \forall\ \tilde{k} \in \overline{1, N_1}$, equal to or smaller than negative d, the middle data set $\tilde{\mathbf{V}}_2$ (N_2 samples) consists of reordered values $\tilde{y}(\tilde{k}) \equiv 0\ \forall\ \tilde{k} \in \overline{1, N_2}$ that just were higher than negative d, but equal to d positive or lower it. The right-hand side data set $\mathbf{V}_3$ (N_3 samples) consists of the reordered in ascending order values $\tilde{y}(\tilde{k})\ \forall\ \tilde{k} \in \overline{1, N_3}$ higher than d. Here $\tilde{k}$ is any integer k rearranged in an ascending order, associated with the reordered values of observations $y(k)\ \forall\ k \in \overline{1, N}$.

Hence, the reordered observations with the positive values higher than d will be concentrated on the right-hand side set, while the reordered observations with the negative values equal $-d$ or smaller than $-d$ on the left-hand side set. The observations of the middle data set of $\tilde{y}(\tilde{k})$ correspond to the intermediate signal $x(k)$, values of which in a non-noisy case acquired the threshold d meaning while passing the deadzone nonlinearity (Fig. 8.1). Thus, for the deadzone alike for the PWL nonlinearity (7.6) there exist three regimes, as well. Now, it could be easy to separate the particles of $\tilde{y}(k)$ by determining the middle data set $\tilde{\mathbf{V}}_2$ samples), if the measurement noise $\upsilon(k)$ is absent. In such a case, the threshold d value can be determined analyzing the graph, where the reordered output-input dependency is shown. However, for

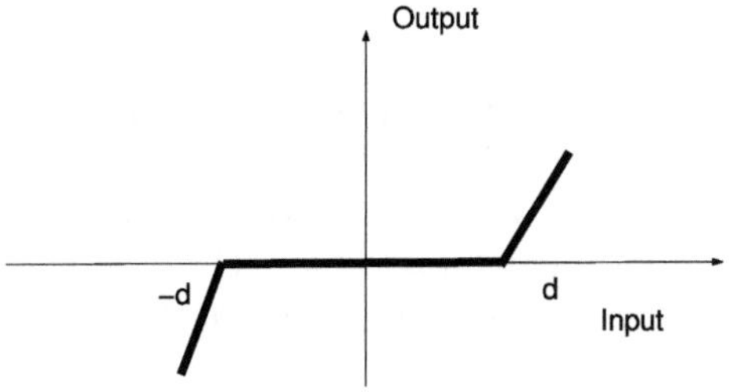

Fig. 8.1 Deadzone nonlinearity $f(\cdot, \eta)$

the deadzone there exists the main difference as compare with the PWL nonlinearity because in such a case, the only left and right data sets $\tilde{\mathbf{V}}_1$, $\tilde{\mathbf{V}}_3$ are informative enough to reconstruct the intermediate signal $x(k)$.

Firstly, parameters ϕ of FIR model are estimated by the ordinary LS as follows:

$$\hat{\phi}_{13} = \left(\tilde{\mathbf{\Pi}}_{13}^T \tilde{\mathbf{\Pi}}_{13}\right)^{-1} \tilde{\mathbf{\Pi}}_{13}^T \tilde{\mathbf{V}}_{13}, \tag{8.12}$$

secondly, the reconstructed version

$$\hat{x}(k) = \hat{\phi}_1 \tilde{u}(k-1) + \hat{\phi}_2 \tilde{u}(k-2) + \ldots + \hat{\phi}_\nu \tilde{u}(k-\nu) \tag{8.13}$$

of the intermediate signal $x(k)$ is determined $\forall\, k \in \overline{m+1, \mathrm{L}}$. In Eq. (8.12):

$$\tilde{\mathbf{V}}_{13} = \begin{bmatrix} \tilde{\mathbf{V}}_1 \\ \tilde{\mathbf{V}}_3 \end{bmatrix}; \tilde{\mathbf{\Pi}}_{13} = \begin{bmatrix} \tilde{\mathbf{\Pi}}_1 \\ \tilde{\mathbf{\Pi}}_3 \end{bmatrix}; \tilde{\phi}_{13} = \begin{bmatrix} \tilde{\phi}_1 \\ \tilde{\phi}_3 \end{bmatrix}. \tag{8.14}$$

$L = N_1 + N_3$. Then, the true values of coefficients (8.6) are replaced with their estimates. Thus, the reconstructed version of the intermediate signal $x(k)$ is obtained in Eq. (8.13), using recursive calculations of $\hat{x}(k)$.

8.3 Parametric Identification of Hammerstein System Having the Deadzone

Assume now that the deadzone nonlinearity precedes linear block of the block-oriented nonlinear system, true output of which is in relation with the true input as follows:

$$x(k) = h_1 u(k-1) + h_2 u(k-2) \ldots + h_{m-1} u(k-m+1) + h_m u(k-m) + e(k) \tag{8.15}$$

$\forall\, k \in \overline{m, m+\mathrm{N}}$.

Rewriting Eq. (8.15) in a matrix form we have

$$\mathbf{X} = \mathbf{\Gamma h} + \mathbf{e}. \tag{8.16}$$

Here

$$\mathbf{X} = \left(\hat{x}(m), \hat{x}(m-1), \ldots, \hat{x}(m+\mathrm{N}-1), \hat{x}(m+\mathrm{N})\right)^T \tag{8.17}$$

is the $(\mathrm{N}+1)\times 1$ vector, consisting of reconstructed samples of the intermediate signal $x(k)$; $\mathbf{e} = (e(m), e(m+1), \ldots, e(m+\mathrm{N}-1), e\,(m+\mathrm{N})))^T$ is the $(\mathrm{N}+1)\times 1$ vector of the sequence of residuals;

$$\mathbf{\Gamma} = \begin{bmatrix} u(m-1) & u(m-2) & \dots & u(2) & u(1) \\ u(m) & u(m-1) & \dots & u(3) & u(2) \\ \vdots & \vdots & \vdots & \vdots & \vdots \\ u(m+\mathrm{N}-2) & u(m+\mathrm{N}-3) & \dots & u(\mathrm{N}+1) & u(\mathrm{N}) \\ u(m+\mathrm{N}-1) & u(m+\mathrm{N}-2) & \dots & u(\mathrm{N}+2) & u(\mathrm{N}+1) \end{bmatrix} \tag{8.18}$$

is the full rank $(\mathrm{N}+1) \times \nu$ regression matrix, consisting only of observations of the input $u(k)$; $\nu = m - 1$,

$$\mathbf{h}^T = (h_1, h_2, \dots, h_{\nu-1}, h_\nu) \tag{8.19}$$

is $\nu \times 1$ vector of unknown parameters of FIR model (8.15).

Next, let us rearrange the output data in the vector $\mathbf{X}$ in an ascending order of their values

$$\tilde{\mathbf{X}} = \left(\tilde{x}(\tilde{k}+1), \tilde{x}(\tilde{k}+2), \dots, \tilde{x}(\mathrm{N}-1), \tilde{x}(\mathrm{N})\right)^T, \tag{8.20}$$

is the $\mathrm{N}\times 1$ vector (8.17), reordered here with the associated rows of the matrix $\mathbf{\Gamma}$ as well. We do that by interchanging equations in the initial matrix (8.18) as follows

$$\tilde{\mathbf{\Gamma}} = \begin{bmatrix} \tilde{u}(\tilde{k}) & \tilde{u}(\tilde{k}) & \dots & \tilde{u}(\tilde{2}) & \tilde{u}(\tilde{1}) \\ \tilde{u}(\tilde{k}+1) & \tilde{u}(\tilde{k}) & \dots & \tilde{u}(\tilde{3}) & \tilde{u}(\tilde{2}) \\ \vdots & \vdots & \vdots & \vdots & \vdots \\ \tilde{u}(\tilde{\mathrm{N}}-2) & \tilde{u}(\tilde{\mathrm{N}}-3) & \dots & \tilde{u}(\tilde{\mathrm{N}}-\nu+2) & \tilde{u}(\tilde{\mathrm{N}}-\nu+1) \\ \tilde{u}(\tilde{\mathrm{N}}-1) & \tilde{u}(\tilde{\mathrm{N}}-2) & \dots & \tilde{u}(\tilde{\mathrm{N}}-\nu+1) & \tilde{u}(\tilde{\mathrm{N}}-\nu) \end{bmatrix}, \tag{8.21}$$

which is the full rank $(\mathrm{N} \times \nu)$ regression matrix, consisting only of observations of the input $u(k)$, reordered according reordered samples of internal output $\tilde{x}(k)$;

Note that values of the reordered $\tilde{x}(k)\ \forall\, k \in \overline{\mathrm{N}_1 + l_1, \mathrm{N}_2 - l_2}$ are determined by data reordering routine given in Sect. 7.5. The vector of the sequence of residuals $\mathbf{e} = (\tilde{e}(\tilde{k}+1), \tilde{e}(\tilde{k}+2), \dots, \tilde{e}(\mathrm{N}-1), \tilde{e}\,(\mathrm{N})))^T$ in (8.16) is reordered automatically as well.

Then, the vector $\mathbf{X}$ and the matrix $\mathbf{\Gamma}$ should be partitioned into three data sets: the left-hand data set $\tilde{\mathbf{X}}_1 = \tilde{\mathbf{\Gamma}}_1\tilde{h}_1$, the middle data set $\tilde{\mathbf{X}}_2 = \tilde{\mathbf{\Gamma}}_2\tilde{h}_2$ and the right-hand data set $\tilde{\mathbf{X}}_3 = \tilde{\mathbf{\Gamma}}_3\tilde{h}_3$ accordingly to the three regimes of the deadzone nonlinearity (8.2). Here $\tilde{\mathbf{X}}_1, \tilde{\mathbf{X}}_2, \tilde{\mathbf{X}}_3$ are $N_1 \times 1$, $N_2 \times 1$ and $N_3 \times 1$ vectors, respectively, while $\tilde{\mathbf{\Gamma}}_1, \tilde{\mathbf{\Gamma}}_2$ and $\tilde{\mathbf{\Gamma}}_3$ are $N_1 \times \nu$, $N_2 \times \nu$ and $N_3 \times \nu$ matrices, correspondingly; $N = N_1 + N_2 + N_3$, moreover $N_1 = N_3$ for simplicity. Thus, initial system (8.16) is reordered into a system

$$\tilde{\mathbf{X}} = \tilde{\mathbf{\Gamma}}\tilde{\mathbf{h}}; \tag{8.22}$$

by interchanging equations in (8.16). In (8.22):

$$\tilde{\mathbf{X}} = \begin{bmatrix} \tilde{\mathbf{X}}_1 \\ \tilde{\mathbf{X}}_2 \\ \tilde{\mathbf{X}}_3 \end{bmatrix}; \tilde{\boldsymbol{\Gamma}} = \begin{bmatrix} \tilde{\boldsymbol{\Gamma}}_1 \\ \tilde{\boldsymbol{\Gamma}}_2 \\ \tilde{\boldsymbol{\Gamma}}_3 \end{bmatrix}; \tilde{\mathbf{h}} = \begin{bmatrix} \tilde{\mathbf{h}}_1 \\ \tilde{\mathbf{h}}_2 \\ \tilde{\mathbf{h}}_3 \end{bmatrix}, \tag{8.23}$$

$$\tilde{\mathbf{h}}_1^T = \left(\tilde{h}_{1,1}, \tilde{h}_{2,1} \ldots, \tilde{h}_{(\nu-1),1}, \tilde{h}_{\nu,1}\right), \tag{8.24}$$

$$\tilde{\mathbf{h}}_2^T = \left(\tilde{h}_{1,2}, \tilde{h}_{2,2} \ldots, \tilde{h}_{(\nu-1),2}, \tilde{h}_{\nu,2}\right), \tag{8.25}$$

$$\tilde{\mathbf{h}}_3^T = \left(\tilde{h}_{1,3}, \tilde{h}_{2,3} \ldots, \tilde{h}_{(\nu-1),3}, \tilde{h}_{\nu,3}\right). \tag{8.26}$$

If the measurement noise $v(k)\ \forall\ k = \overline{1, N}$ is absent, then the left-hand side data set $\tilde{\mathbf{X}}_1$ (N_1 samples) consists of the reordered values $\tilde{x}(\tilde{k})\ \forall\ \tilde{k} \in \overline{1, \mathrm{N}_1}$, equal to or smaller than negative d, the middle data set $\tilde{\mathbf{X}}_2$ (N_2 samples) consists of the reordered cut off values $\tilde{x}(\tilde{k}) \equiv 0\ \forall\ \tilde{k} \in \overline{1, \mathrm{N}_2}$ that just were higher than negative d, but equal to or lower than d. The right-hand side data set $\tilde{\mathbf{X}}_3$ (N_3 samples) consists of the reordered in ascending order values $\tilde{x}(\tilde{k})\ \forall\ \tilde{k} \in \overline{1, \mathrm{N}_3}$ higher than d. Here $\tilde{k}$ is any integer k rearranged in an ascending order, depending on k of the reordered values of observations $x(k)\ \forall\ k \in \overline{1, \mathrm{N}}$. Hence, the reordered observations with the positive values higher than d will be concentrated on the right-hand side set, while the reordered observations with the negative values equal or smaller than $-d$ on the left-hand side set. The observations of the middle data set of $\tilde{x}(\tilde{k})$ correspond to the intermediate signal $x(k)$ values that have become equal zero after the respective values of input signal $u(k)$ was processed by the deadzone nonlinearity. Thus, for the deadzones there exist three regimes, in total. Now, it is easy to separate the particles of $\tilde{x}(k)$ by determining the middle data set $\tilde{\mathbf{X}}_2$ (N_2 samples), if the measurement noise $v(k)$ is absent. In such a case, the d value can be determined approximately, too. The left and right data sets $\tilde{\mathbf{X}}_1, \tilde{\mathbf{X}}_3$ are used to reconstruct the true intermediate signal $x(k)$. Firstly, parameters $\hat{\mathbf{h}}$ of FIR model (8.15) are estimated by

$$\hat{\mathbf{h}} = \left(\tilde{\boldsymbol{\Gamma}}_{13}^T \tilde{\boldsymbol{\Gamma}}_{13}\right)^{-1} \tilde{\boldsymbol{\Gamma}}_{13}^T \tilde{\mathbf{X}}_{13}, \tag{8.27}$$

secondly, the reconstructed version

$$\hat{x}(k+1) = \hat{h}_1 u(k) + \hat{h}_2 u(k-1) + \cdots + \hat{h}_{\nu-1} u(k-\nu+2) + \hat{h}_\nu u(k-\nu+1) \tag{8.28}$$

of the intermediate signal $x(k) \forall\ k \in \overline{\nu, \mathrm{N}}$ is determined. In (8.27):

$$\tilde{\mathbf{X}}_{13} = \begin{bmatrix} \tilde{\mathbf{X}}_1 \\ \tilde{\mathbf{X}}_3 \end{bmatrix}; \tilde{\boldsymbol{\Gamma}}_{13} = \begin{bmatrix} \tilde{\boldsymbol{\Gamma}}_1 \\ \tilde{\boldsymbol{\Gamma}}_3 \end{bmatrix}; \tilde{\mathbf{h}} = \begin{bmatrix} \tilde{\mathbf{h}}_1 \\ \tilde{\mathbf{h}}_3 \end{bmatrix}. \tag{8.29}$$

Thus, the estimate $\hat{x}(k)$ of the intermediate signal $x(k)$ could be extracted using Eq. (8.15), where unknown parameters of FIR model $h_1, h_2 \ldots, h_\nu$ are replaced with their estimates in (8.28). The result of this step is the auxiliary signal $\hat{x}(k)$ that, really, is a reconstructed version of the true intermediate signal $x(k)$, acting between the deadzone nonlinearity and the linear block.

8.4 Numerical Simulation of the Wiener System Having the Deadzone

$N = 100$ data points have been generated without additive process and measurement noises. The sum of sinusoids Eq. (7.31) (see Fig. 8.2a) and white Gaussian noise (Fig. 8.5**a**) are chosen, as training signals to the Wiener system. As input signals of the open-loop flowchart they act on the linear block (7.32) of nonlinear system. The true intermediate signal $x(k)$ $k = \overline{1, N}$, (Figs. 8.2b and 8.5b) of the Wiener system was generated according Eq. (7.5), while the true output signal (Figs. 8.2c and 8.5c) of the Wiener system was realized by Eq. (8.1). The deadzone was described in Eq. (8.2), where the threshold $d = 0.9$. It was determined, analysing the dependence

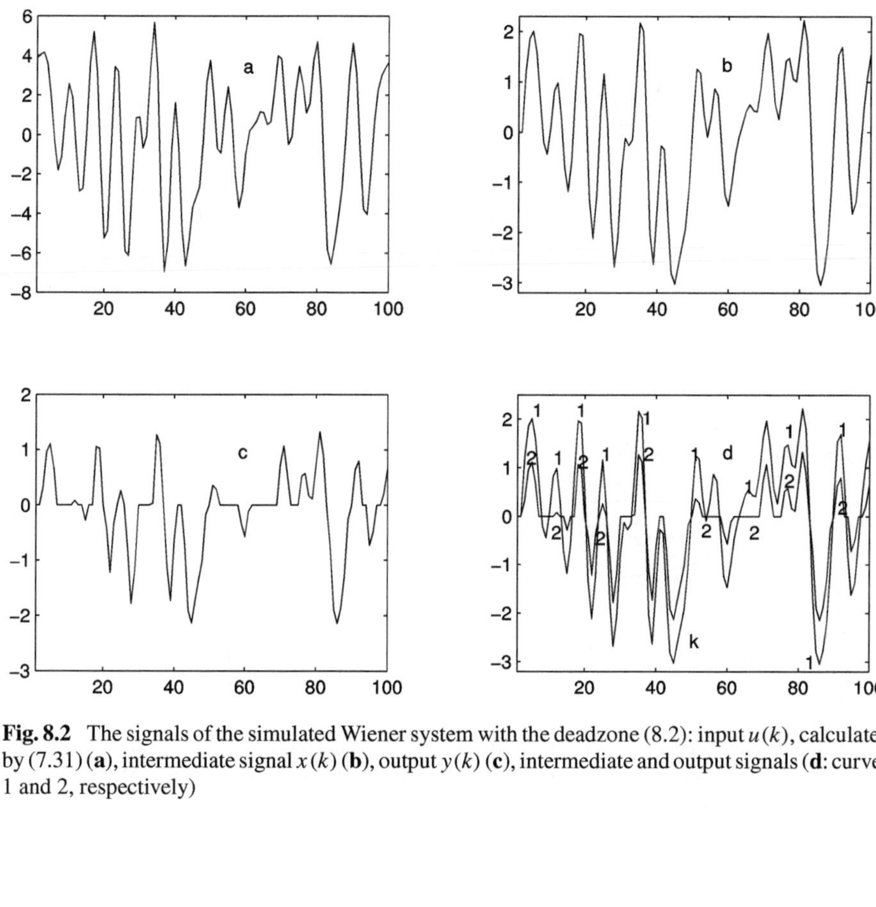

Fig. 8.2 The signals of the simulated Wiener system with the deadzone (8.2): input $u(k)$, calculated by (7.31) (**a**), intermediate signal $x(k)$ (**b**), output $y(k)$ (**c**), intermediate and output signals (**d**: curves 1 and 2, respectively)

of reordered output $y(n)$ values on the reordered in the same way input $x(n)$ values. Such dependences for distinct training signals are given in Figs. 8.3 and 8.4 parts d.

Then, the LS problem (see Eq. (7.13)) was solved, using jointly 30 and 40 rearranged observations of the output Fig. 8.8 (parts b and d, respectively), if training signal is the sum of sinusoids Eq. (7.31) (Fig. 8.2a). 20 and 60 rearranged observations of the output (Fig. 8.9b, d) were applied in a case of white Gaussian noise (Fig. 8.5a). The whole number of FIR filter parameters $\nu = 14$ has been chosen based on the previous results (Tables 7.1 and 7.2), obtained for different order numbers ν of FIR model. The estimate $\hat{x}(k)$ of the intermediate signal $x(k)$ was reconstructed according to (7.21). The reconstructed versions of the intermediate signal $x(k)$ are shown in (Figs. 8.10a and 8.11a). The estimates $\hat{\mathbf{\Theta}}^T = \left(\hat{b}_1, \hat{a}_1\right)$ of parameters of the transfer function $G(q^{-1}, \mathbf{\Theta})$ were calculated in Eq. (7.22), using the observations of the auxiliary signal $\hat{x}(k)$. Afterwards, the estimate $\hat{x}_1(k)$ of the internal signal $x(k)$ was recalculated in Eq. (7.34), using $\hat{b}_1, \hat{a}_1$ and $\hat{x}_1(1) = 0$. In such a case, the estimates $\hat{b}_1, \hat{a}_1$ were chosen being equal true parameters: $b_1 = 0.3, a_1 = 0.5$. The reconstructed versions of the intermediate signal $x(k)$, calculated by Eq. (7.34) are shown in (Figs. 8.10b and 8.11b). It ought to be noted that the accuracy of esti-

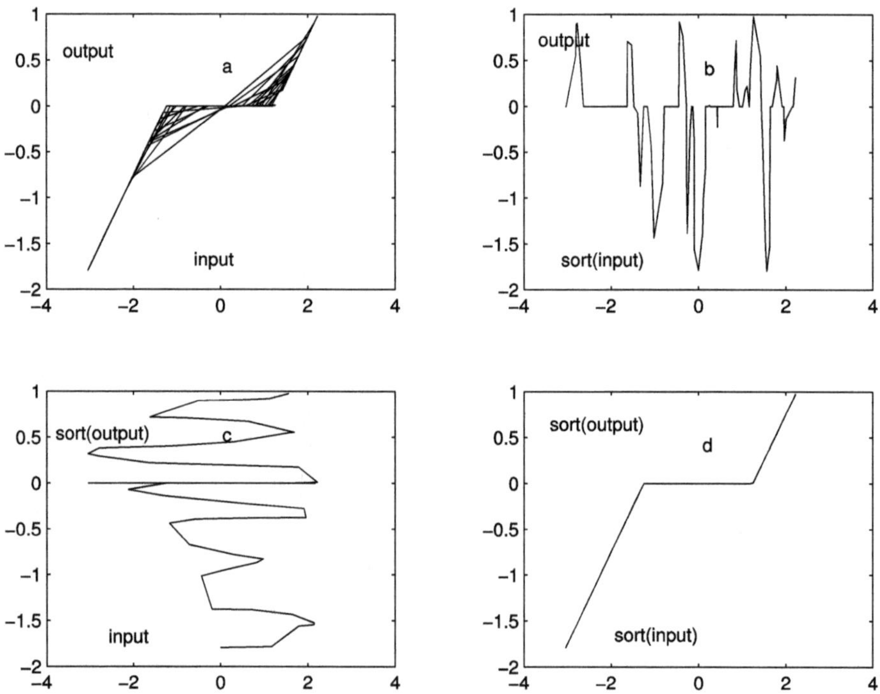

Fig. 8.3 The dependences: of output signal $y(t)$ on the input $x(t)$ (**a**), of $y(t)$ on `sort`$(x(t))$ (**b**), of `sort`$(y(t))$ on $x(t)$ (**c**), and of `sort`$(y(t))$ on `sort`$(x(t))$ (**d**), where `sort` is Matlab routine. Input $x(k)$ is of the form (7.31)

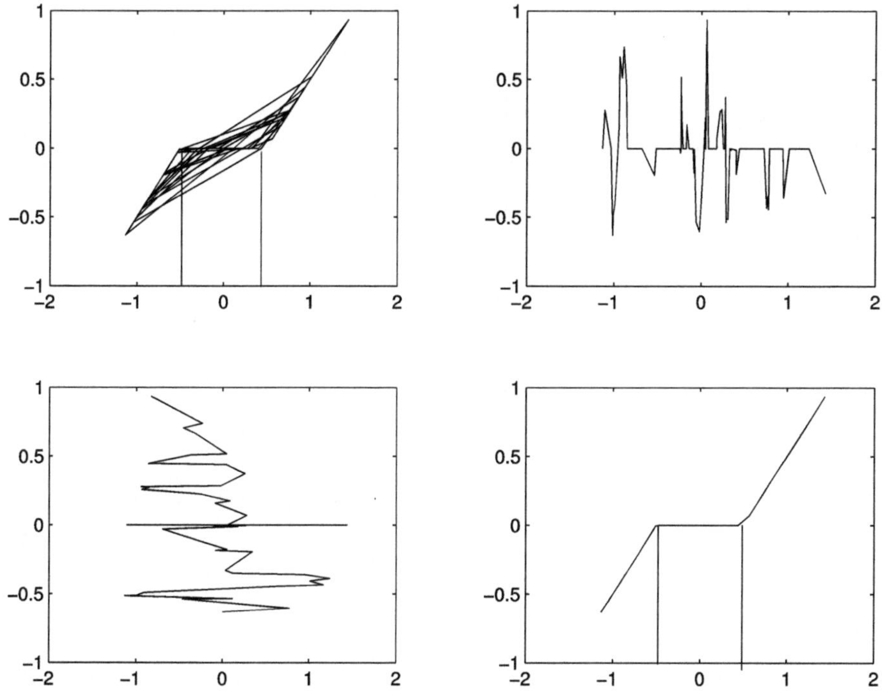

Fig. 8.4 The values, notation and markings are the same as in Fig. 7.6, except that $x(k)$ is white Gaussian noise process realization

mates of the intermediate signal, calculated by formulas (7.21) and (7.34), is more or less similar except for the first 15 observations, when the FIR model (7.21) was used. If $\hat{x}(k)$ has been obtained, then it is simple to separate different particles of observations that belong to the respective side-sets. The estimates of the threshold d were established analysing dependences shown in Figs. 8.3 and 8.4. The values of estimates $\hat{d}$ were less or more equal the true value $d = 0.9$.

8.5 Monte Carlo Simulation

In order to determine how realizations of different process- and measurement noises affect the accuracy of estimation of unknown parameters, we have used the Monte Carlo simulation with 10 sets of data samples, each containing 100 pairs of input-output observations. 10 experiments with the same realization of the process noise $w(k)$ and different realizations of the measurement noise $v(k)$ with different levels of their intensity have been carried out. The intensity of noises was assured choosing respective SNR. For the process noise, $\mathrm{SNR}^{\mathrm{w}}$ was equal to 100, and for the measurement noise, $\mathrm{SNR}^{\mathrm{v}}$: 1, 10, 100. The value of SNR^{w} in each experiment was fixed and

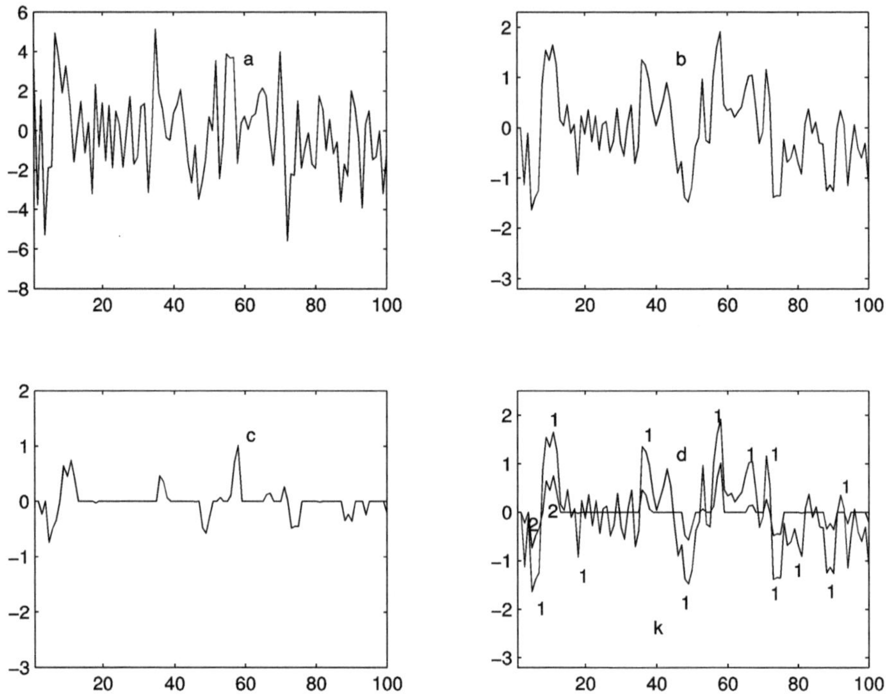

Fig. 8.5 The signals of the simulated Wiener system having the deadzone (8.2): input $u(k)$ is white Gaussian noise (**a**), intermediate signal $x(k)$ (**b**), output $y(k)$ (**c**), intermediate and output signals (**d**: curves 1 and 2, respectively)

was the same, while the values of SNR^{v} were varying. As training signals for all given nonlinearities the periodical signal (7.31) and white Gaussian noise with variance 1 were chosen. In each ith experiment $i = 1, 2, ..., 10$ the estimates of parameters were determined. Averaged values of estimates of the parameters and of the threshold, and their confidence intervals were calculated while the Monte Carlo simulation. In Tables 8.1, 8.2, 8.3, 8.4 and 8.5 are simulation results given. Here for each training signal the averaged estimates of parameters and the threshold d of the simulated Wiener system (Fig. 1.3) with the linear part (7.32) ($b_1 = 0.3$; $a_1 = -0.5$) and the deadzone nonlinearity with their confidence intervals are presented. Analysing values written in Tables 8.1, 8.2, 8.3, 8.4 and 8.5 we noticed that the accuracy of parametric identification of the linear block depends on the FIR model order and on the intensity of measurement noise.

Table 8.1 The dependence of estimates $\hat{b}_1, \hat{a}_1$ of the parameters b_1, a_1 on the number of the FIR model (7.21) parameter ν

Input	Estimates	$\nu = 5$	$\nu = 10$	$\nu = 14$
Periodical	$\hat{b}_1$	0.28	0.29	0.3
Signal (7.31)	$\hat{a}_1$	−0.49	−0.5	−0.5
Gaussian	$\hat{b}_1$	0.29	0.3	0.3
White noise	$\hat{a}_1$	−0.49	−0.49	−0.5

Table 8.2 Averaged estimates of the parameters b_1, a_1 with their confidence intervals. Input: the periodical signal (7.31). SNR^w=100

Input	Estimates	$\mathrm{SNR}^v = 1$	$\mathrm{SNR}^v = 10$	$\mathrm{SNR}^v = 100$
Periodical	$\hat{b}_1$	0.28 ± 0.07	0.3 ± 0.00	0.3 ± 0.00
Signal (7.31)	$\hat{a}_1$	-0.52 ± 0.06	-0.5 ± 0.00	-0.5 ± 0.00
Gaussian	$\hat{b}_1$	0.31 ± 0.04	0.3 ± 0.00	0.3 ± 0.00
White noise	$\hat{a}_1$	-0.39 ± 0.08	-0.5 ± 0.01	-0.5 ± 0.00

Table 8.3 The values and notation are the same as in Table 8.1. Input—the Gaussian white noise

Estimates	$\nu = 5$	$\nu = 10$	$\nu = 14$
$\hat{b}_1$	0.29	0.3	0.3
$\hat{a}_1$	−0.49	−0.49	−0.5
$\hat{a}$	0.96	0.98	1
$-\hat{a}$	−0.97	−0.99	−1

Table 8.4 Averaged estimates of the parameters b_1, a_1 and thresholds $a, -a$ with their confidence intervals. Input: the periodical signal (7.31.) SNR^w=100

Estimates	$\mathrm{SNR}^v = 1$	$\mathrm{SNR}^v = 10$	$\mathrm{SNR}^v = 100$
$\hat{b}_1$	0.28 ± 0.17	0.3 ± 0.08	0.3 ± 0.04
$\hat{a}_1$	-0.52 ± 0.16	-0.5 ± 0.09	-0.5 ± 0.02
$\hat{a}$	0.97 ± 0.28	1 ± 0.12	1 ± 0.07
$-\hat{a}$	-1 ± 0.24	-1 ± 0.13	-1 ± 0.05

Table 8.5 The values and notation are the same as in Table 8.4. Input—the Gaussian white noise

Estimates	$\mathrm{SNR}^v = 1$	$\mathrm{SNR}^v = 10$	$\mathrm{SNR}^v = 100$
$\hat{b}_1$	0.31 ± 0.14	0.3 ± 0.08	0.3 ± 0.03
$\hat{a}_1$	-0.39 ± 0.18	-0.5 ± 0.08	-0.5 ± 0.04
$\hat{a}$	1.03 ± 0.49	0.99 ± 0.74	1 ± 0.12
$-\hat{a}$	-1.21 ± 0.39	-1 ± 0.82	-1 ± 0.09

8.6 Concluding Remarks

The basic idea behind PWL regression is "that if the data follow different linear trends over different regions of the data then we should model the regression function in pieces" (in Internet: online/stat.psu.edu./stat501/lesson /8/8.10/ 2023). The block-oriented nonlinear system could be represented by different threshold regression models, relating with the separate linear model, described by a difference equation. Indeed, the model of the system consists of some blocks or subsystems, among which switchings occur at occasional time moments. For the parametric identification of such systems, the ordinary LS can be applied, if it is known how to recognize observations that belong to different subsystems. Really, observations do not possess distinct signs that could help us to attribute them to different data sets. Therefore, usually a problem of parametric identification of a nonlinear system having a PWL nonlinearity is solved as a problem of the Wiener or Hammerstein systems with a invertible or noninvertible nonlinearities with all well-known consequences beforehand [3]. The nonlinear filtering approach based on the mixed-integer programming method [5] or compound operator decomposition approach (*Vörös* 2010b) could be here used,

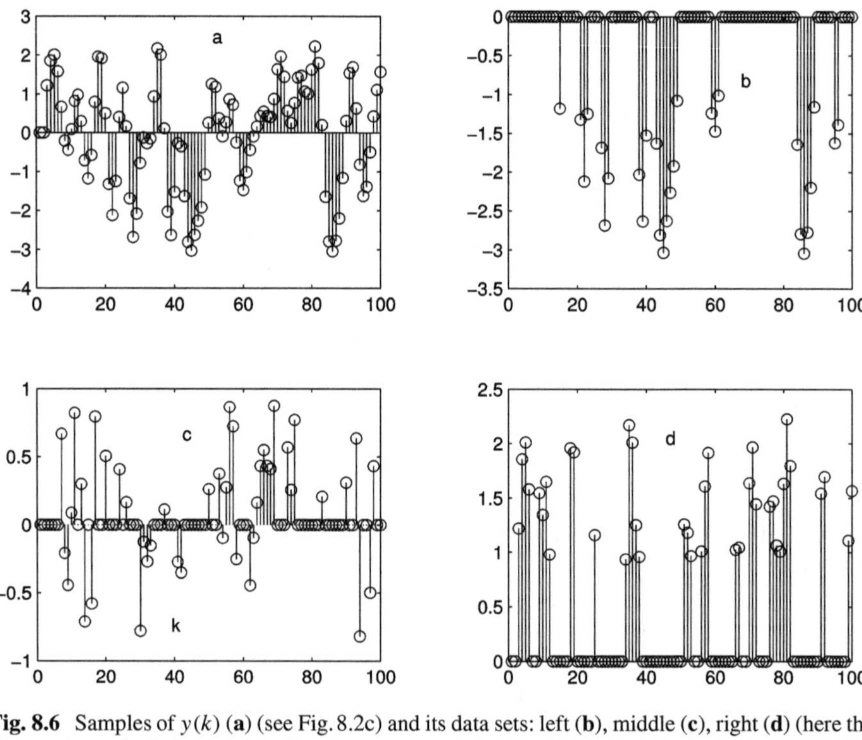

Fig. 8.6 Samples of $y(k)$ (**a**) (see Fig. 8.2c) and its data sets: left (**b**), middle (**c**), right (**d**) (here the observations, that belong to the other data set, are equal to zeros). Input $u(k)$ of the form (7.31) (see Fig. 8.2a)

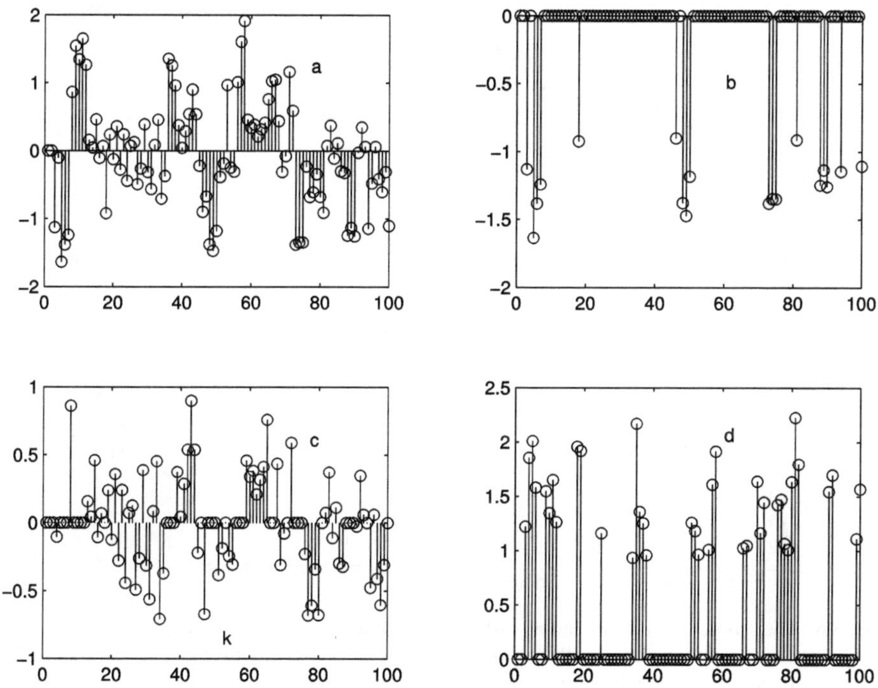

Fig. 8.7 Samples of signal $y(k)$ (**a**) (see Fig. 8.5c) and its data sets: left (**b**), middle (c), right (**d**). Input $u(k)$ is white Gaussian noise (see Fig. 8.5a)

as well. The approach based on the extraction of an unmeasurable internal signal, acting between linear and nonlinear blocks of the Wiener system with deadzone, and avoiding compensators is presented here. The simulation results, shown in Figs. 8.2–8.10 and 8.11 and Tables 8.1–8.4 and 8.5, imply that the iterative technique, based on the output data rearrangement and on the FIR model could be used for the one-shot parametric identification of a Wiener system. Further, the internal signal has been extracted by Eq. (8.13). The accuracy of estimates $\hat{\mathbf{\Theta}}^T(k) = (\hat{b}_1(k), \hat{a}_1(k))$ of true parameters $\mathbf{\Theta}^T(k) = (b_1(k), a_1(k))$ depends on the number of correct observations that were not damaged by the nonlinearity. Note that coefficients $\hat{b}_1(k), \hat{a}_1(k)$ are needed to get the output of the controller.

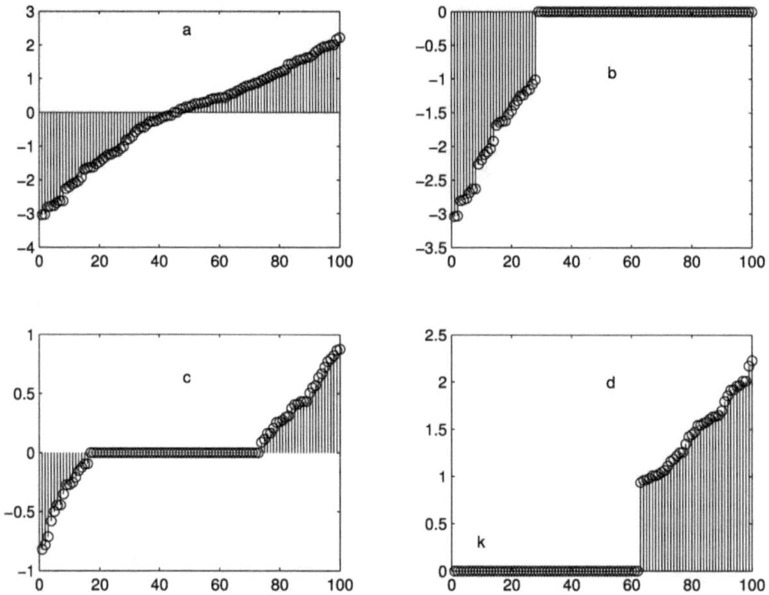

Fig. 8.8 The reordered in an ascending order of their values signal $y(k)$ (**a**) (see Fig. 8.6a) and its rearranged data sets: left (**b**), middle (**c**), right (**d**) (here the observations, that belong to the other data set, are equal to zeros). Input $u(k)$ of the form (7.31) (see Fig. 8.2a)

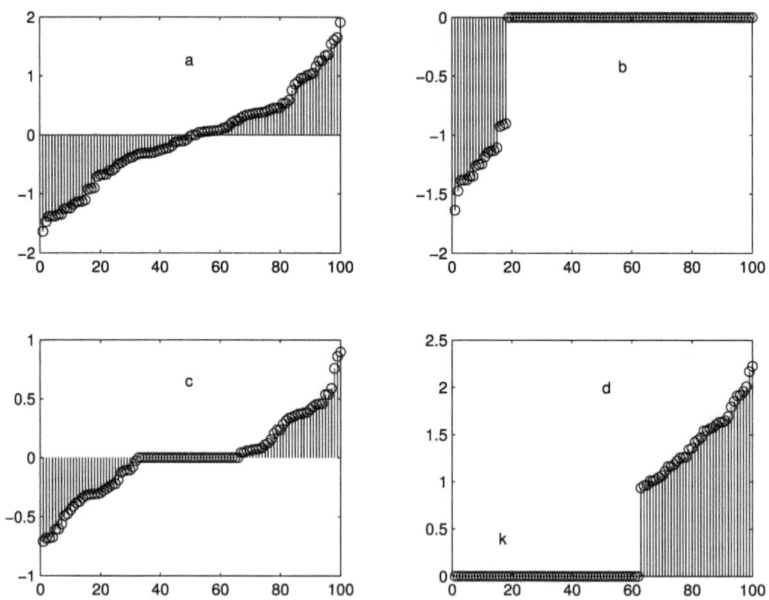

Fig. 8.9 The reordered in an ascending order of their values signal $y(k)$ (**a**) (see Fig. 8.7**c**), and its rearranged data sets: left (**b**), middle (**c**), right (**d**). Input $u(k)$ is white Gaussian noise (see Fig. 8.5a)

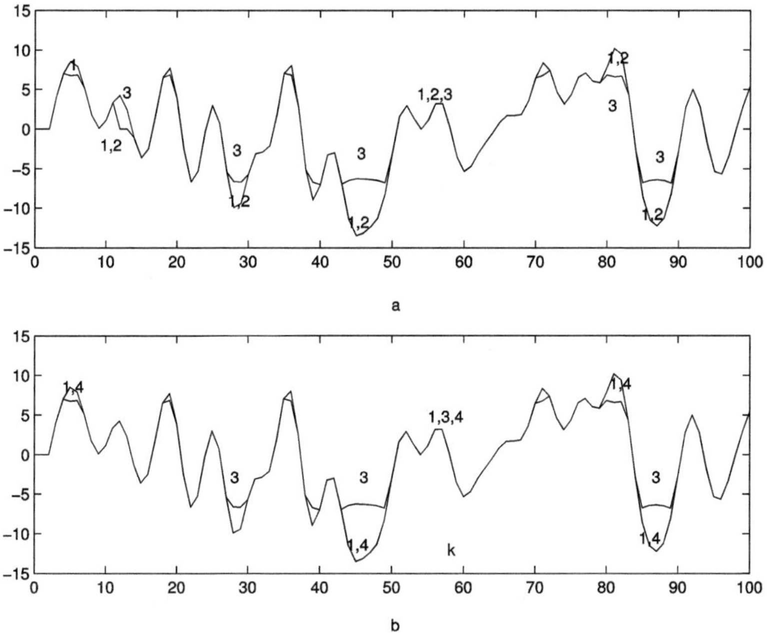

Fig. 8.10 The intermediate signal $x(k)$ (curve 1), the output signal $y(k)$ (curve 3), the reconstructed versions of $x(k)$ (curves 2, 4), calculated using Eq. (7.21) (**a**) and Eq. (7.34) (**b**), Input $u(k)$ of the form (7.31) (see Fig. 8.2a)

The accuracy of parametric identification of the linear block of the Wiener system having the deadzone nonlinearity depends on the intensity of process and measurement noises (see Tables 8.1–8.4 and 8.5). For system, acting in a noisy frame, there appeares multiple local minima of function to be optimized. In such a case, the algorithms based on mixed-integer linear/quadratic programming could be preferable [5].

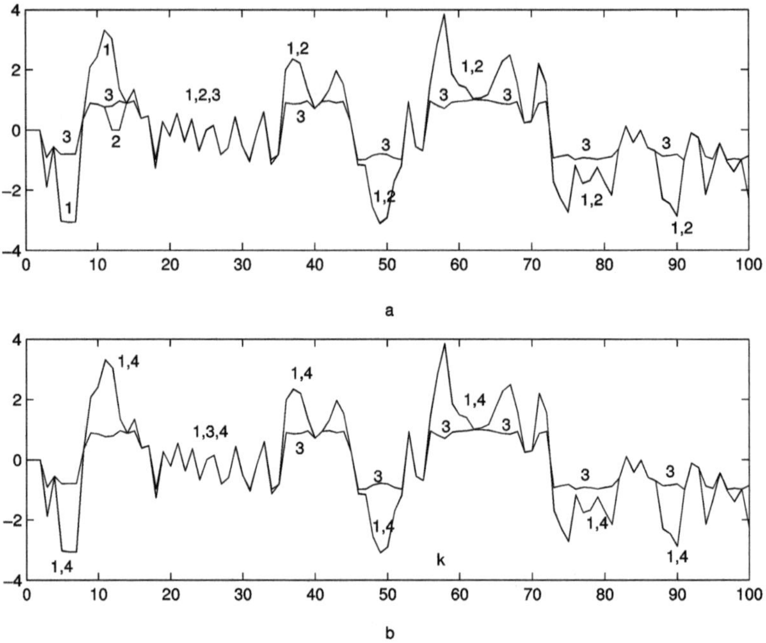

Fig. 8.11 The intermediate signal $x(k)$ (curve 1), the output signal $y(k)$ (curve 3), the reconstructed versions of $x(k)$ (curves 2, 4), calculated using Eq. (7.21) (**a**) and Eq. (7.34) (**b**), Input $u(k)$ is white Gaussian noise (see Fig. 8.5**a**), respectively

References

1. Ljung, L.: System Identification, Theory for the User, 2nd edn. Prentice-Hall PTR (1999)
2. Pupeikis, R., Navakauskas, D., Ljung, L.: Identification of Wiener systems with hard and discontinuous nonlinearities. Technical Report LiTH-ISY-R-2501, Department of Electrical Engineering, Linköping University, SE-581 83 Linköping, Sweden (2003)
3. Pupeikis, R.: Identification of piecewise affine Wiener systems using data partition, Technical Report LiTH-ISY-R-2523, Department of Electrical Engineering, Linköping University, SE-581 83 Linköping, Sweden (2003)
4. Pupeikis, R.: On the Identification of Wiener systems having saturation-like functions with positive slopes. Informatica, LT MII VU Press Vilnius **16**(1), 131–144 (2005)
5. Roll, J.: Local and Piecewise Affine Approach to System Identification, Ph.D thesis No 802. Department of Electrical Engineering, Linköping University, Sweden (2003)
6. Vörös, J.: Parameter identification of Wiener systems with multisegment piecewise-linear nonlinearities. Syst. Control Lett. **56**, 99–105 (2007)

Chapter 9
Controlled Wiener System Simulation

We represent here results of numerical simulation of MV ST control for the Wiener system with the invertible as well as noninvertible PWL nonlinearities. Simulation was performed for an unnoisy and noisy cases.

9.1 Initial Assumptions

At the beginning of the parametrical identification it is useful to consider the assumptions given e.g. in Bobál et al. [1]. It is emphasized *firstly* that the controlled object must be identifiable, which means that it is possible to model it by analysis of measurable values of input and output variables, *secondly*, the model must be selected from the class of so-named parametric models, depending on a finite number of parameters, determination of which is the subject of identification, *thirdly*, a criterion ought to be determined to validate the model to be chosen from the set of models from the given class. For the criterion, it is necessary to fulfil the following conditions:

- it must be suitable measure for detecting the differences between model and object behaviour;
- it must be sensitive to random effects and errors;
- it must be easily computable.

When identifying a given plant, we should follow the preparation procedure described by Bobál et al. [1]:

1. A choice of the most suitable input (exciting) signal, a trade-off between the theoretical optimal excitement and that applied, with respect to the technology to be used. The process of identification can be observed, interrupted, and the input signal can be altered.
2. The data gathered during the experiment can be stored and subsequently processed using various methods with different models, filtered, etc.

R. Pupeikis and K. Kazlauskas, *Advanced Digital Signal Processing Methods for Filtering, Identification, and Nonlinear Systems Control*, Studies in Systems, Decision and Control 638, https://doi.org/10.1007/978-3-032-06740-1_9

3. The model parameters obtained can be tested using other sampled data.
4. The identification experiment can be repeated, possibly with the knowledge gained from previous experiments.
5. Conditions for the lack of bias of the estimates can be tested or verified.

While performing identification for adaptive control, it is necessary to meet the following conditions:

- The data (inputs) are generated by a feedback controller.
- The aim of the controller is to compensate for disturbances and to stabilize the process. These circumstances makes it more difficult to identify process parameters.
- The identification process for adaptive control takes an infinitely long time. It is therefore, not possible to assume constant parameter estimates. Methods to estimate time-varying parameters are essential.
- The identification must be functional with various operating conditions of the plant (at relative steady-state, with disturbances, and during transitions between different states).
- The structure of the identified model (order) cannot usually be changed while running.
- The identification algorithm must be numerically reliable and sufficiently fast.

If the assumptions are met, adaptive control can get successfully.

Training phase. We describe here, in short, how the needed information was obtained for the main system control time. It was assumed that the PWL switched nonlinearity was invertible. Therefore, the Wiener system having invertible three segments PWL nonlinearity (7.30) and linear block (6.52) was chosen for the processing in learning phase. We supposed that the parameters of PWL nonlinearity, as well as that of the linear block were unknown. Therefore, it was necessary to calculate their estimates in oder to substitute them in respective expressions. Thus, to implement the MV ST controller for LTI system correctly, it was needed, firstly, in each current k-th recursion to estimate the parameters b_1, b_2, and a_1, a_2 recursively, and, secondly, to use estimates to determine coefficients of the MV controller (Fig. 9.1). Note that there the closed-loop Wiener system's input-output signals as well as the controller's output values were simulated based on the true and estimated parameter values, respectively.

The whole number of FIR filter (7.8) parameters $\nu = 14$ was chosen. Afterwards, during the learning time an ordinary LS problem (7.13) was solved, processing 50 pairs of rearranged observations. The estimate $\hat{x}(k)$ of the intermediate signal $x(k)$ was determined using Eq. (7.8) where, instead of the true but unknown values $\mathbf{h}$, their estimates $\hat{\mathbf{h}}$ were substituted in (7.21). The auxiliary signal $\hat{x}(k)$ was calculated that is a reconstructed version of the unknown intermediate signal $x(k)$. At the next step the observation samples of the auxiliary signal $\hat{x}(k)$ were used to calculate the estimates $\hat{b}_1, \hat{b}_2, \hat{a}_1, \hat{a}_2$ of the transfer function $G(q^{-1}, \mathbf{\Theta})$ parameters by Eq. (7.22). Afterwards, the estimate $\hat{x}_1(k)$ of the unknown intermediate signal $x(k)$ was recalculated in

$$\hat{x}_1(k) = \hat{b}_1\, u(k-1) + \hat{b}_2\, u(k-2) + \hat{a}_1\hat{x}_1(k-1) + \hat{a}_1\hat{x}_1(k-2), \tag{9.1}$$

$\forall k = \overline{2, 100}$, using $\hat{b}_1, \hat{b}_2, \hat{a}_1, \hat{a}_2$ and $\hat{x}_1(1) = \hat{x}_1(0) = 0$.

In each iteration the current value of the output of the Wiener system $y(k)\ \forall\ k = \overline{1, 1000}$ was filtered by the inverse nonlinear filter (6.67), in which current estimates $\hat{c}_0(k)$, $\hat{c}_1(k)$, $\hat{c}_0'(k)$, $\hat{c}_1'(k)$ were substituted instead of true and unknown parameters $c_0(k)$, $c_1(k)$, $c_0'(k)$, $c_1'(k)$, respectively.

N = 1000 data samples were generated by

$$\check{y}(k) = y(k) + \kappa\varsigma(k), \tag{9.2}$$

adding a measurement noise in order to obtain noisy output values. Here value of $\varsigma(k)$ was initially chosen so that SNR was equal to 50, κ is a coefficient, assuring corresponding SNR value, while a signal is acting in a noisy frame. To determine how realizations of different measurement noise values affect the accuracy of estimation of unknown parameters, we have used the `Monte Carlo` simulation method with 10 data sets, each containing 400 pairs of input-noisy output observation samples. 10 experiments with the different realizations of the measurement noise $n(k) \equiv \varsigma(k)\forall k = \overline{1, 1000}$ have been carried out. Afterwards, the current estimates of signals and parameters were calculated by processing samples of signals $u(k)$, $y(k)$ and $r(k)$. The averaged values of estimates of the parameters and their confidence intervals

$$\Delta = \pm t_{\alpha/2;L-1}\frac{\hat{\sigma}}{\sqrt{L}}. \tag{9.3}$$

were determined, using the `Monte Carlo` simulation method. Here $\hat{\sigma}$ is the estimate of the std; α is the significance level. The value $t_{\alpha/2;L-1}$ is the point of Student's distribution with L-1 degrees of freedom which determines $\alpha/2$ part of the distribution. From Student's distribution table we have $t_{0.025;99} = 1.960$ for $\alpha = 0.05$ and $L = 10$.

9.2 Inverse Nonlinearity Influence While Tracking of the Set-Point Signal

Three experiments, simulated by computer, are described here. Initial conditions of each experiment are given in Table 9.1. Their results are shown in Figs. 9.1, 9.2, 9.3 and 9.4. Initially, N=100 data points were generated without the process and measurement noises $w(k)$, $v(k)$ (Fig. 1.3). The PWL switched nonlinearity was absent during the Experiment 1. Therefore, the data reordering in the learning phase was not used. To implement the MV ST controller for LTI system correctly, it is necessary, firstly, to estimate the parameters b_1, b_2, and a_1, a_2 in each current kth recursion. The LS estimates $\hat{b}_0(k)$, $\hat{b}_1(k)$, $\hat{a}_1(k)$, $\hat{a}_2(k)$ of parameters b_0, b_1, a_1, a_2 of the LTI system

Table 9.1 Characteristics of each experiment in an unnoisy environment

Experiment	Noises	PWL nonlinearity	Learning phase	Figures
1	No	No	Yes	Figure 9.1
2	No	Yes (not fixed)	Yes	Figure 9.2
3	No	Yes	Yes	Figure 9.4

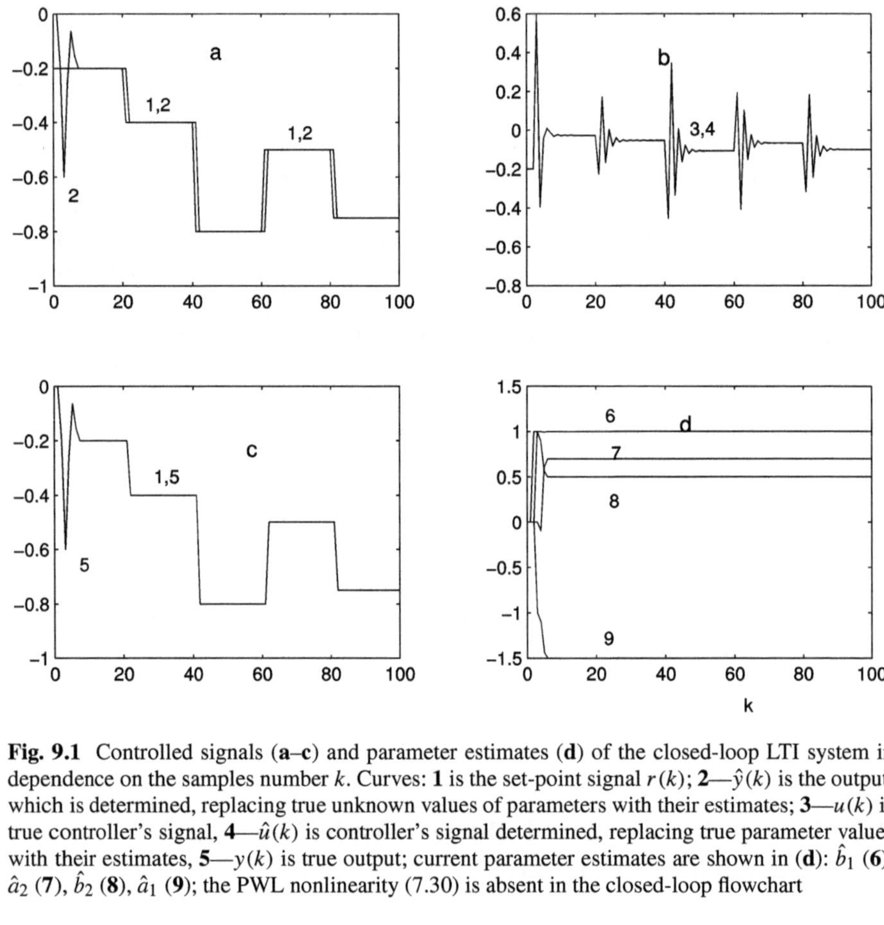

Fig. 9.1 Controlled signals (**a**–**c**) and parameter estimates (**d**) of the closed-loop LTI system in dependence on the samples number k. Curves: **1** is the set-point signal $r(k)$; **2**—$\hat{y}(k)$ is the output, which is determined, replacing true unknown values of parameters with their estimates; **3**—$u(k)$ is true controller's signal, **4**—$\hat{u}(k)$ is controller's signal determined, replacing true parameter values with their estimates, **5**—$y(k)$ is true output; current parameter estimates are shown in (**d**): $\hat{b}_1$ (**6**), $\hat{a}_2$ (**7**), $\hat{b}_2$ (**8**), $\hat{a}_1$ (**9**); the PWL nonlinearity (7.30) is absent in the closed-loop flowchart

were calculated in the learning phase as well. Afterwards, we determine coefficient estimates of the MV controller, applying LS estimates, calculated at last Kth recursion. Note that in all three experiments parameter estimates were left unchangeable during the main phase time, i.e. while control of the Wiener system. The reconstructed version of output $\hat{y}(k)$, was calculated, using abovementioned estimates $\hat{b}_0(K), \hat{b}_1(K), \hat{a}_1(K), \hat{a}_2(K)$ as well. Here K is the maximal value of integer k used to mark the number of last recursion while a parametrical identification.

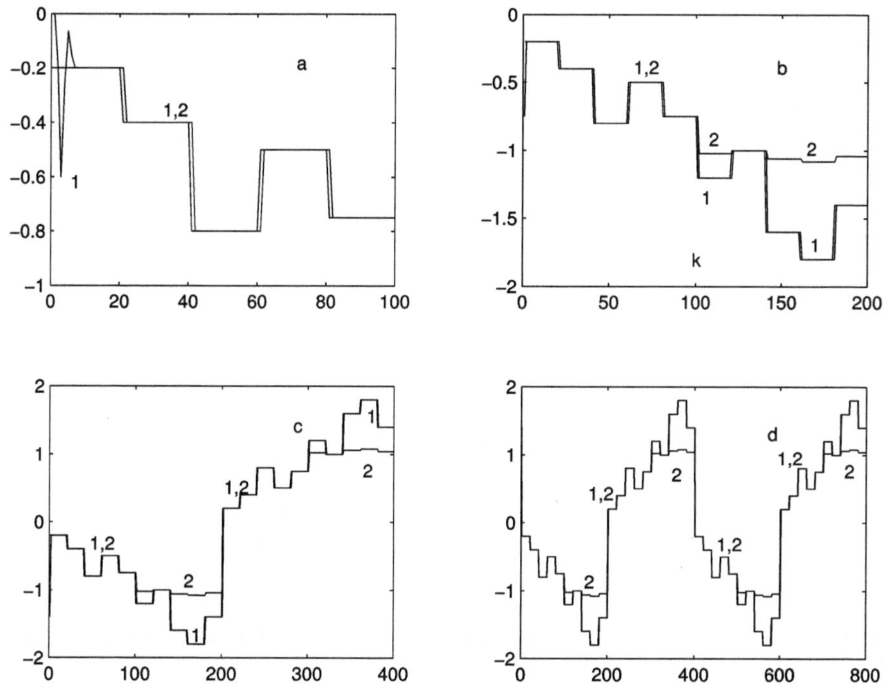

Fig. 9.2 Controlled signals (**a–d**) of the closed-loop Wiener system in an unnoisy frame. Curves: **1**-reference signal $r(k)$, **2**-output $\hat{y}(k)$. PWL nonlinearity is present but not fixed. The inverse nonlinearity (6.66) is absent

It follows from simulation results (Fig. 9.1) that the output of the LTI system $y(k)$ tracks the set-point signal $r(k)$ efficiently (curves 1 and 2). The reconstructed version of $\hat{u}(k)$ (curve 4) corresponds to the true $u(k)$ (curve 3), despite the fact that $\hat{u}(k)$ was obtained using estimates $\hat{b}_1(K)$, $\hat{b}_2(K)$ and $\hat{a}_1(K)$, $\hat{a}_2(K)$. They were substituted instead of respective true values of parameters b_1, b_2 and a_1, a_2 (curve 3). Estimates $\hat{b}_1(k)$, $\hat{b}_2(k)$ of the numerator parameters b_1 and b_2 (curves 6 and 8) as well as estimates $\hat{a}_1(k)$, $\hat{a}_2(k)$ of parameters a_1 and a_2 (curves 7 and 9) of the denominator of system function $G(q^{-1}, \hat{\boldsymbol{\Theta}})$ followed the true parameter values. Reference signal $r(k)$ and both outputs $y(k)$, $\hat{y}(k)$ are shown in Fig. 9.1a, c by curves 1, 2 and 5, respectively. Curves are coincidental. Thus, true and reconstructed outputs $y(k)$ and $\hat{y}(k)$ track the set-point signal $r(k)$ during the training phase, obviously.

Signals of the Wiener system having the PWL switched nonlinearity were processed while Experiments 2 and 3. The PWL nonlinearity (7.30) was introduced into the control flowchart, but was not fixed (Experiment 2). Therefore, the training phase was provided without the data reordering. The output $y(k)$ failed to keep the track of set-point signal $r(k)$ while adaptive servo control process, especially, at the bottom and at the top places of signal $r(k)$ due to the nonlinear effects of the function (7.30) (Fig. 9.2b, c, d). Note that the Wiener system behaves like the LTI system, corrupted

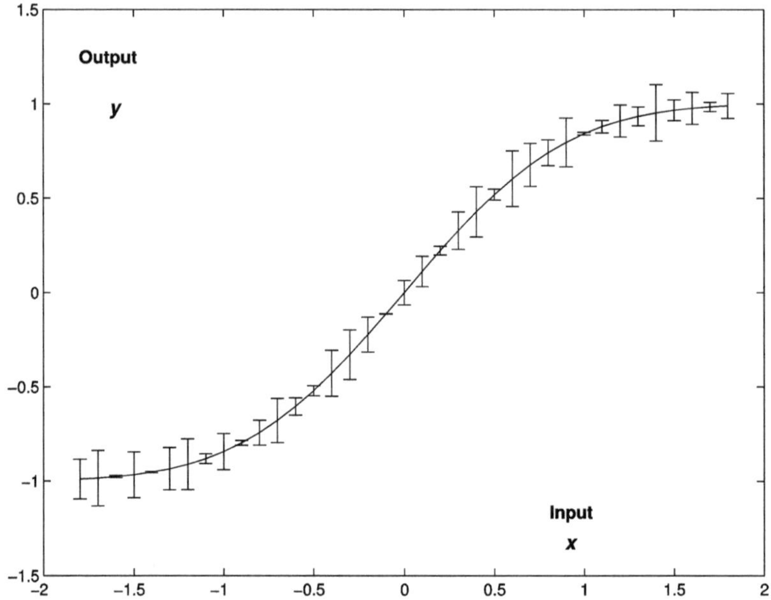

Fig. 9.3 The output $y(k)$ of the PWL nonlinearity with errorbars in dependence of the input $\hat{x}(k)$. The Matlab routine `errorbar` was used

by nonlinear effects, indeed. The current value of the internal signal $x(k)$ (Fig. 9.2c, d) will be cut off if the value of the signal's amplitude is outside the threshold's value of the switched nonlinearity (7.30).

During the training phase the reordered output was subdivided into three segments. We obtain the initial values $\hat{\mathbf{\Theta}}(k) = (0.95, 0.42, -1.37, 0.63)^T$ of parameters b_1, b_2, a_1, a_2, using ordinary LS to process small set of samples $\{u(k), y(k)\}$, containing at most 10 pairs. It was necessary to begin the MV control procedure jointly with the recursive parametric identification of the closed-loop Wiener system. In order to increase the efficiency of MV servo control in Experiment 3, the current value of the output $y(k)$ of Wiener system was updated by passing it through the inverse nonlinear filter (6.66). The threshold's value of the PWL nonlinearity was determined during the training phase of Experiment 3 by joint processing output $y(k)$ and input $\hat{x}(k)$ (the estimate of the intermediate signal $x(k)$) (Fig. 9.3). The values of parameter c_0, c_1 and $c_0^{'}$, $c_1^{'}$ were calculated at the learning phase as well, using left and right data particles, correspondingly, which were found by data reordering. Estimates were substituted in respective places of parameters $c_0, c_1, c_0^{'}, c_1^{'}$ of the inverse nonlinearity (6.66). Estimates $\hat{b}_0(K), \hat{b}_1(K), \hat{a}_1(K), \hat{a}_2(K)$ of parameters b_0, b_1, a_1, a_2 of the LTI system (6.53) and estimates $\hat{c}_0(K_1), \hat{c}_1(K_1), \hat{c}_0^{'}(K_2), \hat{c}_1^{'}(K_2)$ of parameters $c_0, c_1, c_0^{'}, c_1^{'}$ of the switching nonlinearity having the invertible PWL nonlinearity (6.66), were used instead of respective coefficient values of the closed-loop Wiener system. The set of samples of $\hat{x}(k)$, necessary to plot the graph of the PWL nonlinearity, was determined during the learning phase as well. Note that in the training

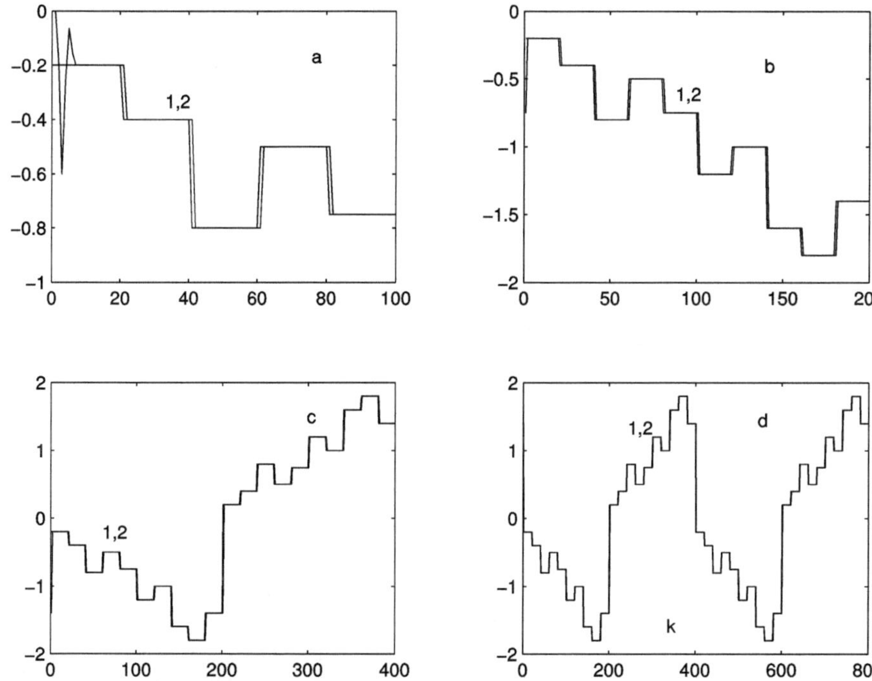

Fig. 9.4 The notation and markings are the same as in Fig. 9.2. PWL nonlinearity is present and fixed. The inverse nonlinearity (6.66) is present

phase the Wiener system was acting in open-loop. Comparison of results, shown in Figs. 9.1, 9.2 and 9.4, corroborate the assertion that it is possible to operate on the output $y(k)$ to track the set-point signal $r(k)$ closely if the nonlinear filter (6.66), inverse to the PWL nonlinearity (7.30), is implemented in the control flowchart.

9.3 Influence of Measurement Noise While Tracking of the Set-Point Signal

Wiener system having linear block (6.52) with the true values of parameters (6.54) and the invertible PWL nonlinearity (6.59) was chosen here for the modelling. MV control law was used to operate on output $y(k)$ to follow the set-point signal $r(k)$. Initial conditions of each experiment are given in Table 9.2.

Firstly, PWL nonlinearity was present but not fixed. The reference signal $r(k)$ (curve 1) and the output $y(k)$ (curve 2) of the closed-loop Wiener system with PWL nonlinearity (6.59), depending on the number of processed samples k are presented in (Fig. 9.5, part a). The indicator of a measurement noise intensity $\text{SNR}^{v} = 10$ was chosen. Note that the switching nonlinearity switched on twice, i.e., when k reached

Table 9.2 Characteristics of each experiment in a presence of measurement noise

Experiment	Noises	PWL nonlinearity	Learning phase	Figures
1	Measurement n.	Yes (not fixed)	Yes	Figure 9.5
2	Measurement n.	Yes	Yes	Figure 9.6

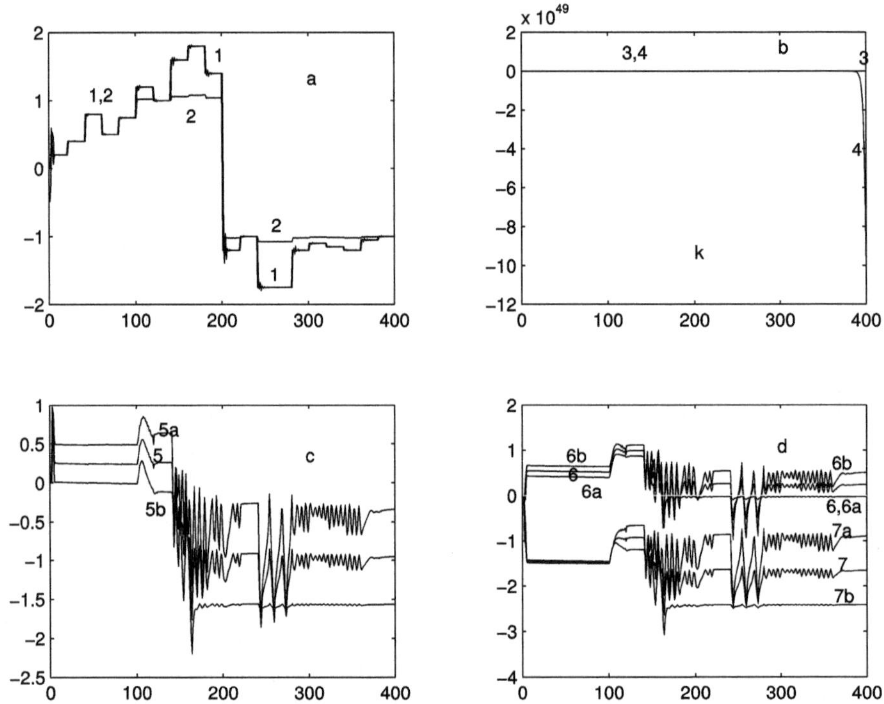

Fig. 9.5 Controlled signals (**a, b**) and parameter estimates (**c, d**) of the closed-loop Wiener system in dependence on the number of samples k. Curves: **1** is set-point signal $r(k)$; **2**—output signal $\hat{y}(k)$; **3**—$u(k)$ (true), **4**—$\hat{u}(k)$, determined by substitution current paraneter estimates instead of true their values; **5**—$\hat{b}_2$; **5**a, **5**b are confidence intervals for current estimate $\hat{b}_2(k)$; $\hat{a}_2$—**7**; **7**a, **7**b are confidence intervals for current estimate $\hat{a}_2(k)$; $\hat{a}_1$—**6**; **6**a, **6**b are confidence intervals for current estimate $\hat{a}_1(k)$. The inverse nonlinearity (6.66) is absent

100 and 200 iterations. Thus, the output $y(k)$ was distorted by PWL nonlinearity (6.59). Therefore, the distorted output $y(k)$ is not coincidental with set-point signal $r(k)$ (see Fig. 9.5, part a). Control signals, true $u(k)$ and estimated $\hat{u}(k)$ (Fig. 9.5, part b), as well as parameter estimates $\hat{\boldsymbol{\Theta}}(k)$ (parts c, d) and true parameter value $\boldsymbol{\Theta}(k)$ are not coincidental, respectively. Note that components b_2, a_1, a_2 of vector parameter $\boldsymbol{\Theta}(k)$ were slightly varying in time, while b_1 value was fixed. In Fig. 9.5 parts c, d estimates of parameters of the transfer function $G(q^{-1}, \boldsymbol{\Theta})$ (curves 5–7) are shown. The middle curve 5 corresponds to the estimate $\hat{b}_2$ of coefficient b_2 in dependence of k. Curves 5a and 5b represent values of the confidence interval (9.3) for each

varying meaning of the estimate $\hat{b}_2$. In part d curves 6 and 7 correspond to varying values of estimates $\hat{a}_1$ and $\hat{a}_2$, respectively. Curves 6a, 6b and 7a, 7b represent their confidence intervals (9.3) depending on k. It is obvious that, parameter estimates $\hat{b}_2, \hat{a}_1, \hat{a}_2$ for distinct k are far away from their true values b_2, a_1, a_2, respectively. Note that estimation results demonstrate nonstability for most data. Besides, it is hard to compare the curves in part b because of relatively large values of the signal $\hat{u}(k)$. To assure the stability of calculations we slightly modify the PLR algorithm's recursive expressions (6.18)–(6.21), using recalculation those parameter values that appeared to be out of area of stability. Nevertheless, the instability distorted the input signal $\hat{u}(k)$ at large k (part b, curve 4).

In the second experiment, the PWL nonlinearity was present and fixed. The restored version $\hat{x}(k)$ of the intermediate signal $x(k)$ was obtained here, as well. The current value of the output of the Wiener system $y(k)$ was filtered by the filter (6.67), where instead of true, but unknown values of parameters c_0, c_1, c_0', c_1' of the switching nonlinearity (6.59) their current estimates $\hat{c}_0(K_1), \hat{c}_1(K_1), \hat{c}_0'(K_2)$ were substituted. In such a case, results of simulation are shown in Fig. 9.6, where notation and markings are the same as in Fig. 9.5. Now, $y(k)$ is fully coincidental with $r(k)$ (see Fig. 9.6a, curves 1, 2). The control signal $\hat{u}(k)$, calculated using parameter estimates $\hat{b}_1(k), \hat{b}_2(k)$ and $\hat{a}_1(k), \hat{a}_2(k)$, approaches true signal $u(k)$ by increasing the

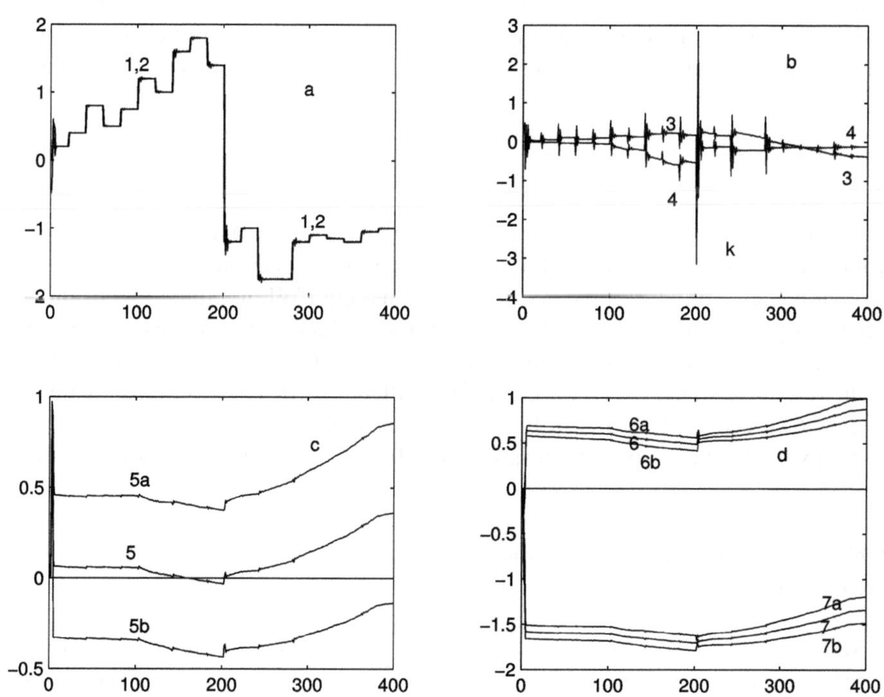

Fig. 9.6 The notation and markings are the same as in Fig. 9.5, with the exception that the inverse nonlinearity (6.67) is used here to eliminate nonlinear distortions

number of samples. Estimates $\hat{b}_1(k)$, $\hat{b}_2(k)$, $\hat{a}_1(k)$, and $\hat{a}_2(k)$ are tracking unknown parameters b_1, b_2, a_1 and a_2 closely enough until boundary of 200 samples. Then, non-accuracy is slowly increasing for all parameter estimates.

We conclude that it is impossible to track signal $r(k)$ if nonlinear distortions created by the PWL nonlinearity are not fully compensated by the inverse filter (6.67), having tuned parameter values.

9.4 Influence of Noises While Tracking of the Set-Point Signal

Here we represent results of three experiments, each of which is executed in a noisy environment, i.e. in a presence of process and measurement noises. Characteristics of abovementioned experiments are given in Table 9.3. In each recursion the current value of the output of the Wiener system $y(k)\ \forall\ k = \overline{1, 1000}$ was filtered by the inverse nonlinear filter (6.67), in which unknown true values of parameters $c_0(k)$, $c_1(k)$, $c_0'(k)$, $c_1'(k)$ were replaced by their estimates $\hat{c}_0(k)$, $\hat{c}_1(k)$, $\hat{c}_0'(k)$, $\hat{c}_1'(k)$, respectively. The behavior of the system was controlled by MV control law, using ST controller. Adaptive controller coefficients were recalculated using values of estimates $\hat{b}_1(k)$, $\hat{b}_2(k)$ and $\hat{a}_1(k)$, $\hat{a}_2(k)$ of slowly linearly time-varying parameters $b_2(k)$ and $a_1(k)$, $a_2(k)$, correspondingly. Then, the current value of the control signal, based on the values of the reference signal $r(k)$ and that of noisy output $y(k)$ is as follows

$$u(k) = \frac{p_w\{\hat{a}_1(k)y(k) + \hat{a}_2(k)y(k-1) - \hat{b}_2(k)u(k-2)\} + g_w r(k)}{p_w\hat{b}_1(k) + q}, \tag{9.4}$$

where p_w, g_w and q are scalar coefficients. If $p_w = g_w = 1$ and $q = 0$ the controller becomes MV controller that will be used in this section. In Fig. 9.7, signals and parameter estimates of the Wiener system having PWL nonlinearity are shown, when noisy frame has $\text{SNR}^w = \text{SNR}^v = 10$. Figure 9.8 represents results, obtained for the same SNR^w=10 and increased $\text{SNR}^v = 20$. In part a curves 1 and 2 correspond to the reference signal $r(k)$ and output $\hat{y}(k)$, filtered by inverse nonlinear switching filter (6.67). Curves 1, 2, 3, 4 in Fig. 9.7c correspond to $\hat{c}_0(k)$, $\hat{c}_1(k)$, $\hat{c'}_0(k)$, $\hat{c'}_1(k)$, respectively. Curves in part d are as follows: $1(\hat{b}_1(k))$, $2(\hat{b}_2(k))$, $3(\hat{a}_1(k))$, $4(\hat{a}_2(k))$.

Table 9.3 Characteristics of each experiment in a noisy frame

Experiment	SNR^v	SNR^n	Inverse PWL nonlinearity	Learning phase	Figures
1-MVC	10	10	Yes	No	Figure 9.7
2-MVC	10	20	Yes	No	Figure 9.8
3-GMVC	10	20	Yes	No	Figure 9.9

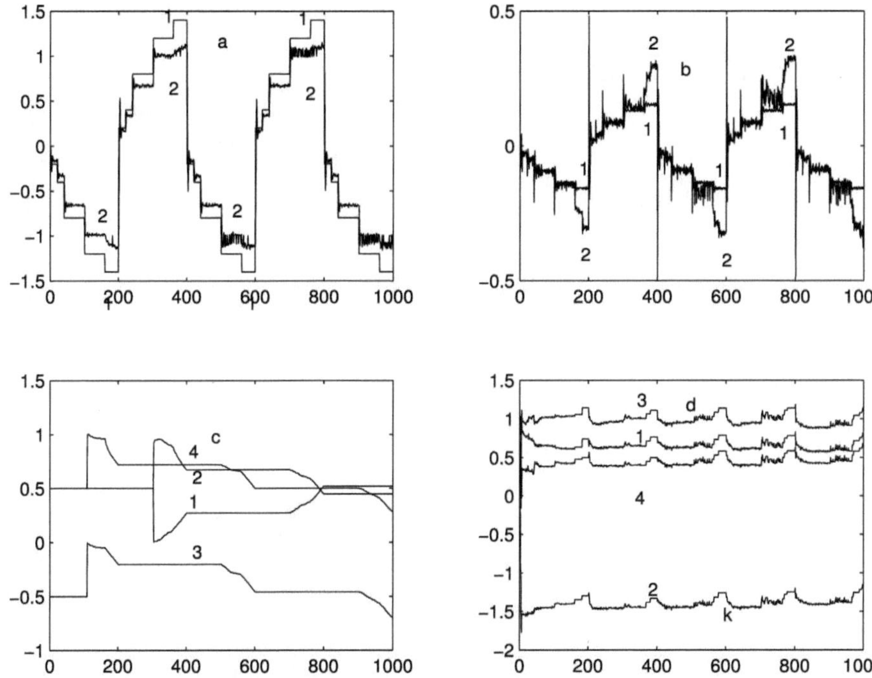

Fig. 9.7 Signals (**a**, **b**) and parameter estimates (**c**, **d**) of the closed-loop Wiener system in a noisy frame. Curves in part a: **1**—reference signal $r(k)$, **2**—output $\hat{y}(k)$; curves in b: **1**—true input $u(k)$ of the form (6.64) with $p = g = 1$ and $q = 0$; **2** corresponds to the input $u(k)$ of form (6.64), in which true parameters $\mathbf{b}_1$, $\mathbf{b}_2$, $\mathbf{a}_1$ and $\mathbf{a}_2$ were replaced by their estimates; curves in **c**: $\hat{c}_0(k)$ **1**, $\hat{c}_1(k)$ **2**, $\hat{c}'_0(k)$ **3**, $\hat{c}'_1(k)$ **4**; curves in (**d**): $\hat{b}_1(k)$ 1, $\hat{b}_2(k)$ 2, $\hat{a}_1(k)$ **3**, $\hat{a}_2(k)$ **4**; the inverse nonlinearity (6.67) is used here; $\mathrm{SNR}^w = \mathrm{SNR}^v = 10$

Analysing deviations of curves 1, 2 in Figs. 9.7a and 9.8a one can state that the accuracy of estimates $\hat{c}_0(k)$, $\hat{c}_1(k)$, $\hat{c'}_0(k)$, $\hat{c'}_1(k)$ (curves 1–4, (part c)), is not sufficient to adjust the output signal $\hat{y}(k)$ properly. Therefore, there exists the clearly visible discrepancy between the set-point signal 1 and the reconstructed output 2. The accuracy of estimates can not be higher, if the GMV control with the internal signal extraction would extend to the MV servo control (Fig. 9.9).

9.5 Minimum Variance Control of the Wiener System with the Deadzone Nonlinearity

The linear block of the nonlinear Wiener system is given by the common LTI system transfer function (7.1). The true values of transfer function $G(q^{-1}, \boldsymbol{\Theta})$ parameters in (7.3) are [2]: $b_1 = 1,\ b_2 = 0.5,\ a_1 = -1.5,\ a_2 = 0.7$. The output of the Wiener

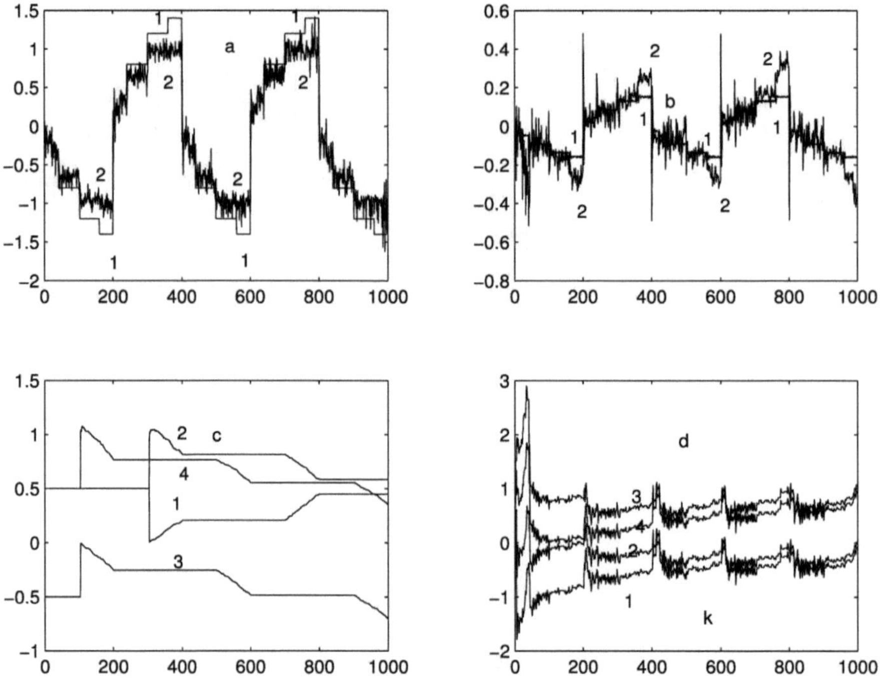

Fig. 9.8 The notation and markings are the same as in Fig. 9.7; $\text{SNR}^w = 10$, $\text{SNR}^v = 20$

system $y(k)$ is generated by (7.4), where the deadzone (7.7) threshold size d is 0.5. It is estimated in advance. The GMV controller for LTI system is given by

$$u(k) = \frac{\hat{a}_1 y(k) + \hat{a}_2 y(k-1) - \hat{b}_2 u(k-1) + r(k)}{\hat{b}_1(k) + q}, \tag{9.5}$$

with $q = 2.5$. To implement the ST GMV controller correctly, it is necessary, firstly, at each current recursive operation to estimate the LTI system parameters b_1, b_2, a_1, a_2 by the recursive expressions (6.18)–(6.21), and, secondly, to substitute their estimated values in Eq. (9.5). Here the values $y(k)$ and $y(k-1)$ are changed by the current values $\hat{x}(k)$ and $\hat{x}(k-1)$, respectively. Then, $N = 2000$ samples have been generated with additive process and measurement noises, using relationships:

$$\kappa(k) = x(k) + \lambda_1 \nu(k), \;\; y(k) = y(k) + \lambda_2 \nu(k) \; \forall k = \overline{1, N}. \tag{9.6}$$

The values of λ_1, λ_2 were chosen so that SNRs were equal 10 for both noises, acting in the flowchart. Finally, the current estimate $\hat{u}(k)$ was determined in Eq. (9.5). Note that in previous chapter it is shown, how to determine current reconstructed value of intermediate signal $x(k)$, using the FIR model.

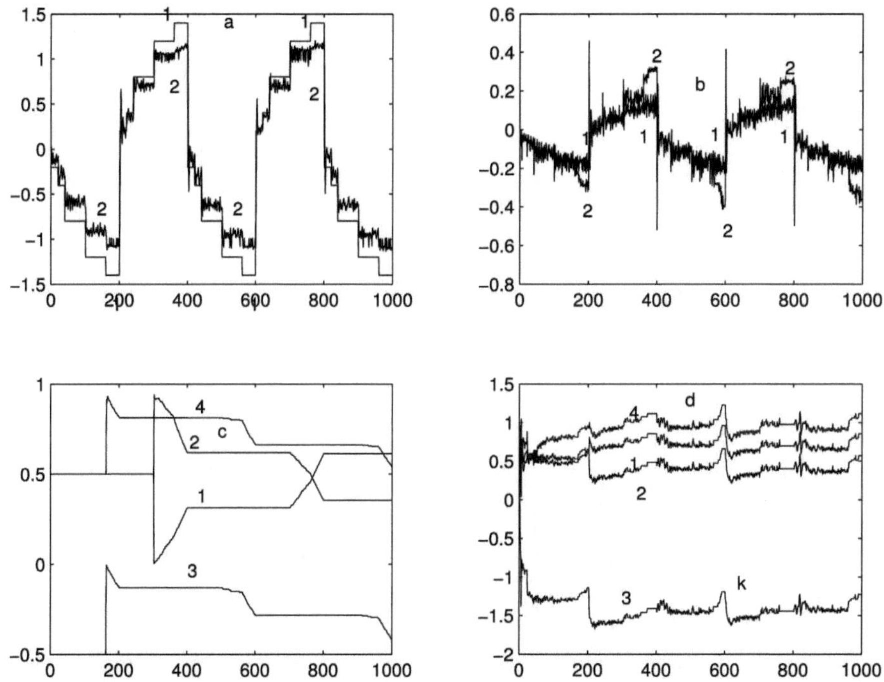

Fig. 9.9 The notation, marking and values are the same as in Fig. 9.8; the GMV control law was used here with $p = q = g = 1$ in recursive equation (9.4)

Simulation results imply that the iterative technique, based on the FIR model (7.8), could be used for the recursive parametrical identification of the Wiener system, if the deadzone inverse

$$f^{-1}(x(k), \eta) = \begin{cases} y(k) - 0.5 & \text{if } y(k) > 0, \\ 0 & \text{if } y(k) = 0, \\ y(k) + 0.5 & \text{if } y(k) < 0 \end{cases} \tag{9.7}$$

containing the threshold $\hat{d} = d = 0.5$ was employed (Fig. 9.11).

Simulation results, obtained in the noisy frame, are shown in Figs. 9.10 and 9.11. The linear block parametric identification results of the Wiener system Fig. 9.11d are obviously better than that shown in Fig. 9.10d. In such a case, the parametric identification results, i.e. estimates of coefficients of the numerator of the transfer function of the linear block of the Wiener system, shown in Fig. 9.11d, are more accurate than that given in Fig. 9.10d. Besides, the value of $\hat{b}_2$ is there unexpected negative. Note that the true value of b_2 is 0.5. In general, the quality of STC of the Wiener system using the inverse of the deadzone nonlinearity has significantly increased in respect of the GMV control without the deadzone inverse Fig. 9.10d.

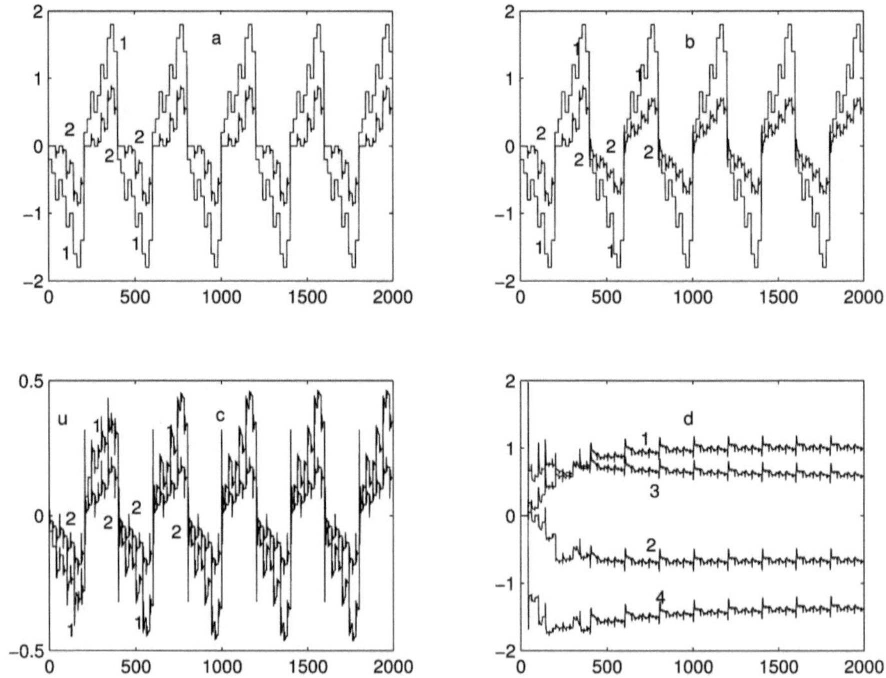

Fig. 9.10 GMV STC of the Wiener system having the deadzone. Curves: 1—set-point signal $r(k)$ (parts **a**, **b**); 2—cut-off and noisy output signal $y(k)$ (part a); 2—estimate $\hat{x}(k)$ of the intermediate signal $x(k)$ (part b); 1, 2—true input $u(k)$ and its estimate $\hat{u}(k)$ (part **c**), respectively. Coefficient estimates vector $\hat{\boldsymbol{\Theta}}^T = (\hat{b}_1, \hat{b}_2, \hat{a}_1, \hat{a}_2)$ (part **d**). In (part **d**): 1 $(\hat{b}_1)$, $2(\hat{b}_2)$, $3(\hat{a}_1)$, $4(\hat{a}_2)$. The inverse deadzone nonlinearity (9.7) is not used here

9.6 Control of the Wiener System with the Noninvertible Piecewise-Linear Nonlinearity

Two different types of MV control could be used for nonlinear systems with distinct PWL nonlinearities: adaptive control in a case of an invertible nonlinearity and non-adaptive one for both types of nonlinearities. During the adaptive control linear block parameter estimates are used to calculate the controller parameter sequentially. The processing time of both algorithms fit into the same sampling period T_s. It should be chosen carefully because sometimes the instability appears (Åström et al. 1984), while the sampling of linear continuous-time equations. A similar behavior can be found in discrete models of nonlinear systems.

The non-adaptive control type may be used in the absence of the inverse function of a PWL nonlinearity. It is realized by the one-shot controller tuning, during which are processed reordered samples of each segment of PWL nonlinearity. Recursive parametric identification and controller parameter calculation routines are connected with each other only in off-line regime while the training time. Once the training time

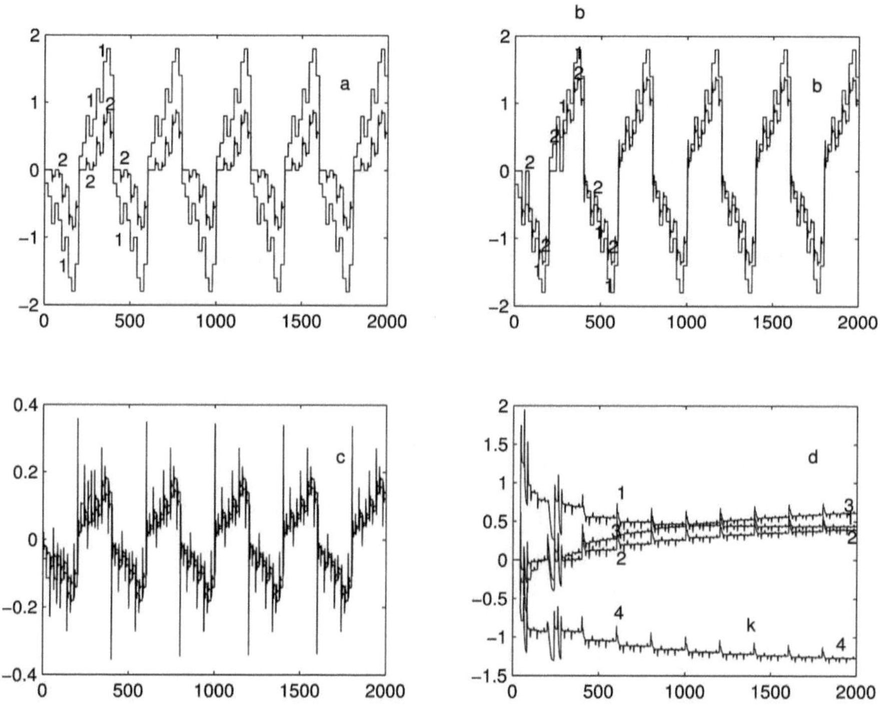

Fig. 9.11 Values and markings are the same as in Fig. 9.10. The inverse deadzone nonlinearity (9.7) is applied here

is over, the parametric identification is finished until the next connection appears. The estimates of unknown linear regression parameters can be calculated by processing respective sets of the rearranged output and associated input observation samples. The unnoisy intermediate signal is fully reconstructed in the training phase.

Note that here the numerical simulation components and visual aids used in Sect. 7.3 were applied. Thus, the true intermediate signal $x(k)$ $k = \overline{1, \mathrm{N}}$, of the Wiener system (Figs. 7.14b and 7.15b) is given by (7.5) with additive process noise $w(k) \equiv 0$ and $\mathrm{N} = 100$. The true output signal (Figs. 7.14c and 7.15c) is described by (7.39) and by (7.40) with the sum of sinusoids (7.31) (Fig. 7.14a) and white Gaussian noise (Fig. 7.15a), respectively, as inputs to the linear block (7.32).

The LTI system is parameterizeted by the FIR model (7.8), which coefficient (7.12) estimates (7.14) were determined by LS method (7.13). The estimate $\hat{x}(k)$ of the intermediate signal $x(k)$ could be determined using Eq. (7.8) where, instead the true values β, their estimates $\hat{\beta}$ are substituted. Thus, the estimate $\hat{x}(k)$ $\forall\, k = \overline{1, M}$ is obtained by (7.21). Here M is the total number of samples of the signal to be processed.

Estimates of parameters c_1, d_1 and c_0, d_0 are calculated in Eqs. (7.27) and (7.28), respectively. In such a case, the rearranged observations of $\hat{x}(k)$ and $y(k)$ were

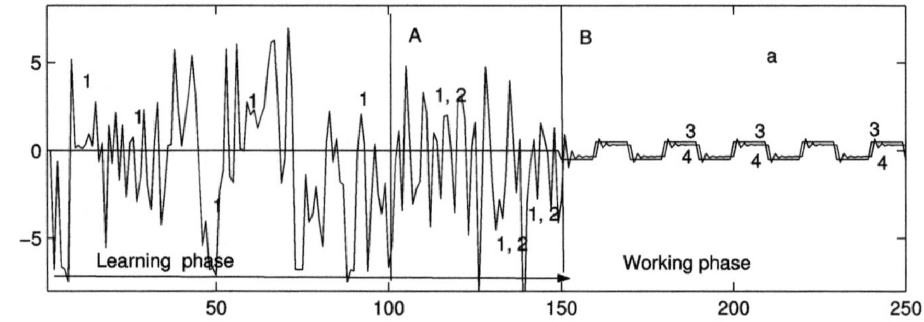

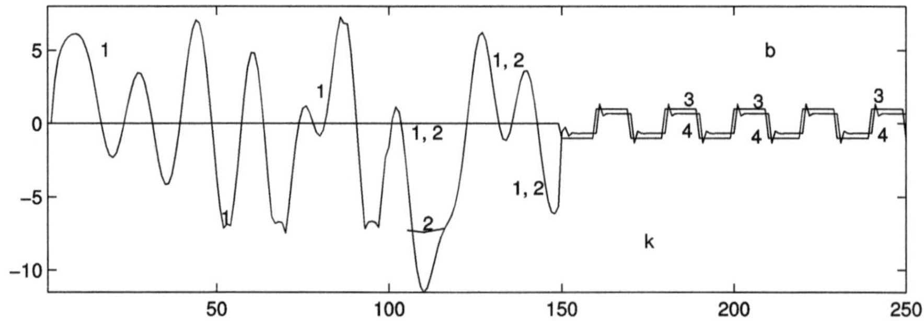

Fig. 9.12 Training and working phases of the Wiener system with non-invertible nonlinearity (7.39) in dependence on processed observation samples. Signals: unnoisy output $y(k)$ (curve 1), intermediate signal $\hat{x}$ (curve 2), set-point signal $r(k)$ (curve 3) and controlled output (curve 4). Input: white Gaussian noise (part a) and the sum of periodical signals with random phases (part b)

substituted in formulas (7.27) and the estimates of c_1 and d_1 were determined: $\hat{c}_1 = \hat{d}_1 = 0.1$. Estimates $\hat{c}_0$ and $\hat{d}_0$ were calculated by (7.28). Their values are also coincidental with the values of true coefficients: $\hat{c}_0 = -7.6$, while $\hat{d}_0 = 7.6$ for the non-invertible nonlinearity (7.39) and $\hat{c}_0 = -1.1$, while $\hat{d}_0 = 1.1$ for the non-invertible nonlinearity (7.40), respectively. It should be noted that $N_1 = 14$, $N_3 = 8$ for the periodical signal (7.31) (Fig. 7.14a), and $N_1 = 32$, $N_3 = 28$ for the Gaussian white noise (Fig. 7.15a) were used to calculate the estimates $\hat{c}_0, \hat{c}_1, \hat{d}_0, \hat{d}_1$, respectively. The values of estimates $\hat{a}$ were equal to the true values: a=7.5 and a=1, correspondingly. The learning and working phases of the Wiener system are shown in Figs. 9.12 and 9.13. Training phase lasts time in which fit 150 observations approximately, while working phase requires 100 samples additionally. Initial 100 observations are used for the parametrical identification of FIR model having 14 coefficients. The intermediate signal $\hat{x}(k)$ is coincidental with the FIR model's output. Set-point signals were changed by rectangular pulse sequences, when training phases were finished. Nevertheless, both outputs tracked new reference signals accuratly enough (except of in Fig. 9.13a).

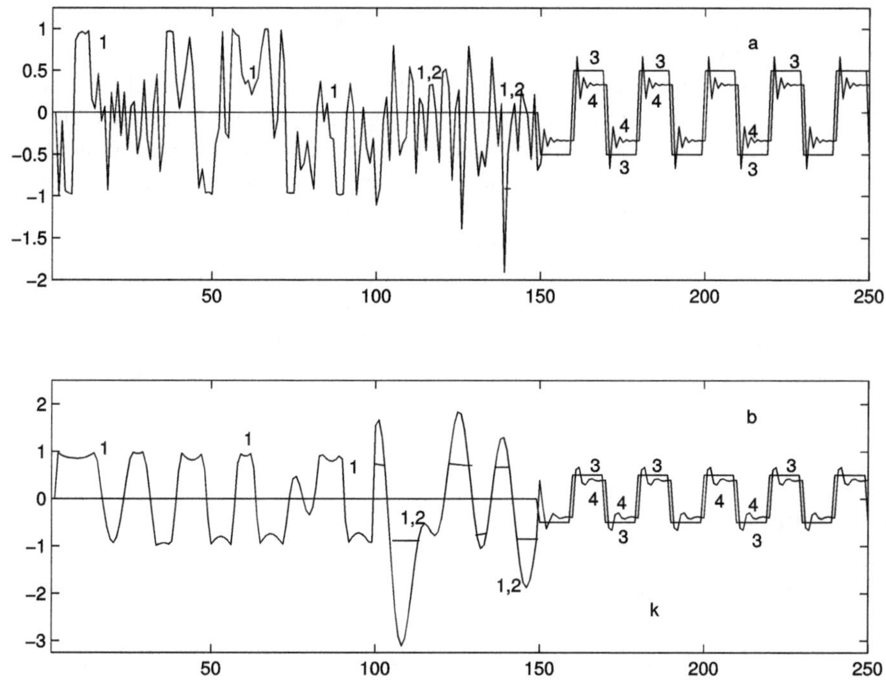

Fig. 9.13 Training and working phases of the Wiener system with non-invertible nonlinearity (7.40) in dependence on processed observations. The notation, marking and values are the same as in Fig. 9.12

9.7 Conclusions

Various results of numerical simulation, including that of Monte Carlo, prove the efficiency of the proposed approach for the MV control of Wiener systems. Comparing the simulation results, presented in Figs. 9.1–9.8 and 9.9, we can assert that in an unnoisy frame the MV ST control of the Wiener system having the PWL nonlinearity is efficient: firstly, when the extraction of the internal signal $x(k)$, based on the samples reordering is used to find the initial values of algorithms and parameters (Figs. 9.1–9.5 and 9.6), and, secondly, if the current parameter estimates $\hat{c}_0(k)$, $\hat{c}_1(k)$, $\hat{c'}_0(k)$, $\hat{c'}_1(k)$, and $\hat{b}_1(k)$, $\hat{b}_2(k)$, $\hat{a}_1(k)$, $\hat{a}_2(k)$ based on the initial values, approach true parameters $c_0(k)$, $c_1(k)$, $c'_0(k)$, $c'_1(k)$, and b_1, b_2, a_1, a_2 closely enough. The efficiency does not fail strongly even in the noisy frame, where both noises $w(k)$ and $v(k)$ $\forall\, k \in \overline{1, N}$ are present Figs. 9.7, 9.8 and 9.9. However, the estimates $\hat{b}_1$, $\hat{b}_2$, $\hat{a}_1$, $\hat{a}_2$ and $\hat{c}_0$, $\hat{c}_1$, $\hat{d}_0$, $\hat{d}_1$ do not approach their true values in the sense of mean squares, even for largest number of observations k, applied to estimate the parameters b_1, b_2, a_1, a_2 and c_0, c_1, d_0, d_1 by noisy signal samples (Figs. 9.7, 9.8 and 9.9). We guess that in such a case there appears more local extremums of minimized function used to calculate unknown parameters in a noisy environment. Distinct combination of parameters led us to the various minimal values of abovementioned function, which do not differ significantly.

It is known [1] that two types of adaptive MV control could be used for nonlinear systems having distinct nonlinearities. During the first control type the recursive parameter estimates, given by the ordinary RLS, are provided for calculation of the controller parameter sequentially. The system and controller parameter estimates calculation process is flowing, in fact, simultaneously all the control time through. The second control type realizes one-shot controller tuning, when recursive parametrical identification and controller parameter calculation routines are connected only during the training phase. Note that the second control type is preferable for the Wiener system having the non-invertible PWL nonlinearity. In such a case, the data reordering approach and the reconstruction of the intermediate signal based FIR model are important.

References

1. Bobál, V., Böhm, J., Fessl, J., Machaček, J.: Digital Self-Tuning Controllers. Springer, Berlin (2005)
2. Kazlauskas, K., Pupeikis, R.: Self-tuning control of linear system followed by deadzones. TEM J. Tech. Educ. Manag. Inform. Novi Pazar: Assoc. Inform. Commun. Technol. Educ. Sci. **3**(1), 3–7 (2014). http://www.temjournal.com/documents/vol3no1/TemJournalFebruary201437.pdf

Appendix A
Consistent Parametric Identification

In this section, the parameter estimation problem is described. Parameter estimates are obtained by the well-known prediction error method. Conditions for estimates consistency are used to the PWL Wiener model parameter estimates. It is shown by Hagenblad (1999) that under additional assumptions on the noise and the true system the approximative predictor still yields a consistent estimate. The output of the open-loop Wiener model (Fig. 1.3) can be written according to Hagenblad (1999)

$$y(k) = f(G(q^{-1}, \boldsymbol{\Theta})u(k) + v(k, \boldsymbol{\eta}) + e(k). \tag{A.1}$$

The formal definition of the predictor $\hat{y}(k, \boldsymbol{\Theta}, \boldsymbol{\eta})$, which depends not only on time $k \equiv kT_s$ but also on parameter vectors $\boldsymbol{\Theta}$ and $\boldsymbol{\eta}$, is then

$$\hat{y}(k, \boldsymbol{\Theta}, \boldsymbol{\eta}) = \mathcal{E}(y(k) \mid \mathbf{Z}^{k-1}, \boldsymbol{\Theta}, \boldsymbol{\eta}) \tag{A.2}$$

In (A.2): $\mathcal{E}$ is the symbol of the conditional expected value, $\mathbf{Z}^{k-1} = \{u(1), y(1), u(2), y(2), ..., u(k), y(k)\}$ is the set of data up to time k.

The prediction error estimate is the parameters vector $\boldsymbol{\Theta}$ and $\boldsymbol{\eta}$, minimizing $V_N(\boldsymbol{\Theta}, \eta)$ as follows (Ljung 1978; Hagenblad 1999):

$$(\hat{\boldsymbol{\Theta}}, \hat{\boldsymbol{\eta}}) = \underset{(\boldsymbol{\Theta}, \boldsymbol{\eta})}{\texttt{argmin}}\, V_N(\boldsymbol{\Theta}, \boldsymbol{\eta}), \tag{A.3}$$

where

$$V_N(\boldsymbol{\Theta}, \boldsymbol{\eta}) = \frac{1}{N}\sum_{k=1}^{N}(y(k) - \hat{y}(k, \boldsymbol{\Theta}, \boldsymbol{\eta}))^2 = \frac{1}{N}\sum_{k=1}^{N}\varepsilon^2(k, \boldsymbol{\Theta}, \boldsymbol{\eta}), \tag{A.4}$$

N is the number of samples, $\varepsilon(k, \boldsymbol{\Theta}, \boldsymbol{\eta})$ is the prediction error.

R. Pupeikis and K. Kazlauskas, *Advanced Digital Signal Processing Methods for Filtering, Identification, and Nonlinear Systems Control*, Studies in Systems, Decision and Control 638, https://doi.org/10.1007/978-3-032-06740-1

The criterion $V_N(\boldsymbol{\Theta}, \eta)$ achieves the smallest value if the predicted $\hat{y}(k, \boldsymbol{\Theta}, \eta)$ and the measured output $y(k)$ samples are approximately coincidental $\forall k = \overline{1, N}$.

Definition 1 (*Consistency*) Suppose that the true system is described by the parameters $\boldsymbol{\Theta}_0$ and η_0. Let $\boldsymbol{\Theta}_N$ and η_N denote the estimates obtained from a data set with N data. The estimate is said to be consistent if $\boldsymbol{\Theta}_N \longrightarrow \boldsymbol{\Theta}_0$ and $\eta_N \longrightarrow \eta_0$ when $N \longrightarrow \infty$. Here $\boldsymbol{\Theta}_0$ and η_0 are true parameter vectors of system linear block and of nonlinear one, respectively.

Let us assume that the minimizing argument $\hat{\boldsymbol{\Theta}}$ of the prediction error criterion is:

$$\hat{\boldsymbol{\Theta}}_N = \underset{\boldsymbol{\Theta}}{\texttt{argmin}} \frac{1}{N} \sum_{k=1}^{N} (y(k) - \hat{y}(k, \boldsymbol{\Theta}))^2. \tag{A.5}$$

The convergence problem has been divided. Firstly, it is shown that $\hat{\boldsymbol{\Theta}}$ converges to the minimizing argument of a deterministic criterion, and, secondly, that the true parameters also minimize that criterion. Under some assumptions on the noise and on the true system, Hagenblad (1999) shows consistency for a less complicated, approximate predictor.

Assumption 1 Measurement noise $\{e(k)\}$ is uncorrelated with old data, $\mathcal{E}(e(k) \mid \mathbf{Z}^{k-1}) = 0$. The predictor then can be written as

$$\hat{y}(k, \Theta, \eta) = \mathcal{E}\left\{f(G(q, \Theta)u(k) + v(k), \eta) \mid \mathbf{Z}^{k-1}\right\}. \tag{A.6}$$

This assumption is, e.g., fulfilled if the measurement noise is white.

Assumption 2 $f(G(q, \boldsymbol{\Theta}_0)u(k) + v(k), \eta_0) = f(G(q, \boldsymbol{\Theta}_0)u(k), \eta_0) \ + \mathcal{E}(\tilde{f}\mathbf{Z}^{k-1}, v(k)) \mid \mathbf{Z}^{k-1}) = 0$.

In this case the predictor according Hagenblad (1999) can be simplified to

$$\hat{y}_s(k, \boldsymbol{\Theta}, \eta) = \mathcal{E}(f(G(q, \boldsymbol{\Theta})u(k), \eta) \mid \mathbf{Z}^{k-1} = f(G(q, \boldsymbol{\Theta}_0)u(k), \eta). \tag{A.7}$$

This is equivalent to saying that the process noise can be transformed to additive noise on the output, which is uncorrelated with past data.

Assumption 2 is fulfilled if f is linear and $\{v(k)\}$ is uncorrelated with past data, or if there is no process noise. The consistency can not be assured if the Assumption 2 is not valid.

It is emphasized by Hagenblad (1999) that together, they will be sufficient to ensure that $\hat{y}(k, \boldsymbol{\Theta}) = \{\hat{y}(k, \boldsymbol{\Theta}_0)$ implies $\boldsymbol{\Theta} = \boldsymbol{\Theta}_0$ (except for a constant gain that can be arbitrarily distributed between the LTI and nonlinear systems if intermediate signal $\{x(k)\}$ is not measured. Additionaly, four different conditions are formulated by Hagenblad (1999):

C1. The linear model structure has to be globally identifiable;
C2. The input data has to be informative enough;
C3. The nonlinearity must be invertible;

C4. The input to the nonlinearity $\{x(k)\}_{k=1}^{N}$ must be dense in the support of f, when N tends to infinity.

Note that a linear dynamic system is a special case of the Wiener model, therefore the first two conditions for linear system must be satisfied. The third condition deals with the coupling between the linear and the nonlinear system, while the last condition is needed to guarantee that the nonlinear system can be identified.

It is globally identifiable if it is globally identifiable at almost all $\boldsymbol{\Theta}^*$.

Examples of globally identifiable structures are ARX models, and output error models where the numerator and denominator polynomials have no common factors (Hagenblad 1999).

Appendix B
Examples of the Use of DSP Methods for Modelling, Filtering, Identification, and Control

By understanding where the proposed methods and algorithms can be applied the reader can better appreciate their importance and efficiency.

Examples in which the layered polynomial structures can be applied:
Polynomial filters are applied in real-time control systems where the dynamics of the system and noise are not known;

They can be used to estimate the state of the system or identify its parameters when dealing with nonlinear systems;

Polynomial type filters are used in navigation systems to process data from various sensors, including inertial systems and distance measurements;

They are advantageous in closed-loop systems like transmitter power control, offering efficient prediction of polynomial-like signals;

Polynomial filters are used in multichannel approaches to improve the spectral flatness of acoustic transfer functions in a room, helping to reduce variability in sound quality;

Polynomial structures are used in various fields, including finance, science, and marketing to identify patterns, and predict outcomes;

Polynomial structures can be used to model the spread of infections diseases and predict the number of cases over time, also can be used in image processing techniques to improve the clarity and accuracy of diagnostic images;

Polynomial-type structures play a significant role in various areas of chemistry, particularly in chemical gaph theory and chemical kinetics. They are used to represent and analyze chemical structures, predict properties and model reaction mechanizm.

Examples in which high speed least mean squares adaptive filtering can be applied:
Noise canceling headphones use LMS filters to remove background noise such as hum of an engine, from a desired audio signal;

R. Pupeikis and K. Kazlauskas, *Advanced Digital Signal Processing Methods for Filtering, Identification, and Nonlinear Systems Control*, Studies in Systems, Decision and Control 638, https://doi.org/10.1007/978-3-032-06740-1

In telecommunications, LMS help to remove echo from voice signals, improving call quality in situations like longe-distance calls or speakerphone conversations;

In communication systems, LMS filters can compensate for the distortion introduced by a communication channel, such as multi-path or inter symbolic interference;

LMS filters can be used to model or identify an unknown system by observing its input and output signals and adjusting the filter coefficients to best match the system's behavior;

The LMS algorithm can be used to model and control the blood preassure, adapting to changes in a patient's condition and ensuring optimal pressure regulation;

In mobile phones, LMS filters help ensure reliable data transmission by compensating for channel distortion;

In control systems, LMS filters can be used to identify the dynamic characteristics of a plant(the system being controlled).

Examples of LPTV systems in which our method can be applied:
Human vocal tract is a time-varying system because the shape of vocal organs changes constantly, affecting a transfer function of the system;

Multirate sampled data control systems involve sampling and processing signals at different rates, which introduces time-varying parameters into the system equations;

Periodically switched resistors whose value changes periodically can be modelled as an LPTV systems;

Rotating systems like wind turbines exhibit LPTV characteristics due to the changing stiffness mand inertia as the blade rotates.

Examples in which iterative parametric identification algorithm can be applied:
Iterative methods are applied to identify parameters in systems where measurements are taken at different sampling rates (dual-rates systems);

The characteristics of speech signals can change rapidly, requiring adaptive AR models for accurate analysis and synthesis;

Stock prices, exchange rates often exhibit non-stationary behavior, making adaptive AR filters useful;

Temperature, air quality can be affected by seasonal changes and other factors, making adaptive modelling necessary.

Examples of approaches that result in piecewise affine systems are (Roll 2003):
Neural networks with piecewise affine perceptrons;

Chua's canonical representation and hinging hyperplanes;

Self-exciting threshold autoregressive models for time-series analysis;

Special-purpose methods for physical applications;

Some function approximation approaches.

Examples of real-time applications in which PWA systems with PWL nonlinearities can be used:
In control systems when there are physical limits on components or caracteristics such as in a tank that can get full or empty;

In most control systems, e.g. in electro-mechanical ones, where so named *hard-nonlinearities*, such as friction, deadzone, saturation, preload, backlash, and hysteresis can occur;

In real-time switched control systems such as chemical or nuclear reactors and partly or fully-autonomous remote-controlled robots or electricity energy systems where rules for the suply and use are established in advance;

In designing digital control systems, while determining the type of uncertainties appearing in the plant to be controlled and inducing negative effects that might sometimes lead to an unstable control, e.g. nonnormal noise due to occasional failures of a signal acquisition device degrade the performance of the control;

In unmanned remote working vehicles that are linked to automated steering functions under circumstances;

In controlling a robot while making the output to follow a reference trajectory.

MIX
Papier aus verantwortungsvollen Quellen
Paper from responsible sources
FSC® C105338

If you have any concerns about our products,
you can contact us on
ProductSafety@springernature.com

In case Publisher is established outside the EU,
the EU authorized representative is:
Springer Nature Customer Service Center GmbH
Europaplatz 3, 69115 Heidelberg, Germany

Printed by Libri Plureos GmbH
in Hamburg, Germany